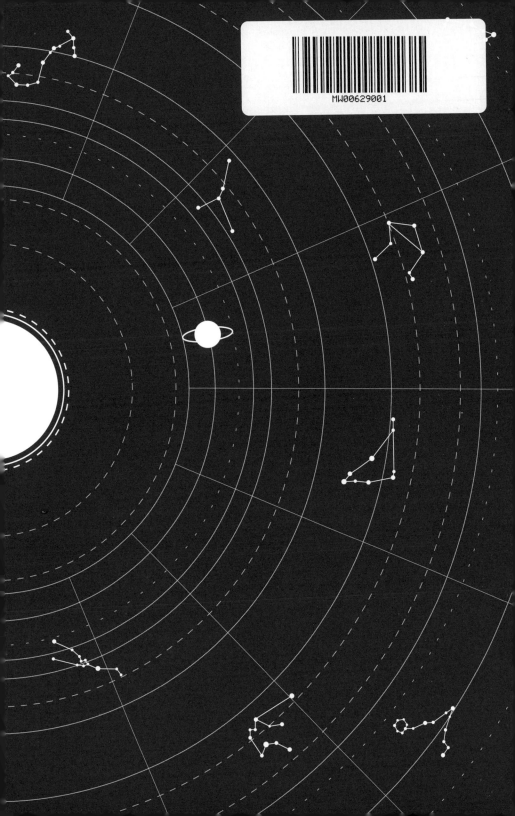

THE
ASTRONOMY
PUZZLE
BOOK

THE ASTRONOMY PUZZLE BOOK

ROYAL OBSERVATORY GREENWICH

PUZZLES BY

DR GARETH MOORE

HODDER &
STOUGHTON

First published in Great Britain in 2021 by Hodder & Stoughton
An Hachette UK company

1

Copyright © National Maritime Museum 2021
Puzzles © Gareth Moore

Illustrations on page 75 © Peter Liddiard, Sudden Impact Media

A CIP catalogue record for this title is available from the British Library

Hardback ISBN 9781529322835
eBook ISBN 9781529322842

Designed by Nicky Barneby, Barneby Ltd
Typeset in 10/15pt Avenir Book by Barneby Ltd

Reproduction by Alta London UK

Printed and bound in Italy by L.E.G.O Spa

Hodder & Stoughton policy is to use papers that are natural, renewable
and recyclable products and made from wood grown in sustainable forests.
The logging and manufacturing processes are expected to conform to
the environmental regulations of the country of origin.

Hodder & Stoughton Ltd
Carmelite House
50 Victoria Embankment
London EC4Y 0DZ

www.rmg.co.uk
www.hodder.co.uk

CONTENTS

INTRODUCTION

Space has the power to inspire and fascinate all of us on Earth, and the history of astronomy has been one of solving puzzles, one after another, as astronomers and scientists have attempted to understand the Universe's many mysteries. The Royal Observatory, in Greenwich, London, occupies an important place in that story, as it encouraged astronomical innovation for hundreds of years and came to define time and place for the world. The Observatory's rich history and collections have provided the inspiration for this book of brain-teasing astronomical puzzles.

A BRIEF HISTORY OF
ROYAL OBSERVATORY GREENWICH

The Royal Observatory was founded in 1675 when King Charles II appointed an astronomer to chart the stars and produce astronomical data to aid navigation, an important task in an age of long-distance ocean voyages and maritime trade. John Flamsteed was appointed as the first Astronomer Royal and Sir Christopher Wren designed the original Observatory building on top of the hill in Greenwich Park. Over the following centuries, ten successive Astronomers Royal expanded the scope of the Observatory's work as it encouraged and led developments in astronomy, navigation and accurate timekeeping. In the late nineteenth century, Greenwich Mean Time became the international time standard, and the Greenwich Meridian was selected as Longitude 0° 0′ 0″, the Prime Meridian of the World.

After the working observatory was moved away from London in the 1940s, the historic buildings were opened to the public in 1960. Today they are part of Royal Museums Greenwich and welcome thousands of visitors each year to discover the Observatory's rich history, walk in the footsteps of the Astronomers Royal and stand on the Prime Meridian. The Astronomy team continue to encourage understanding of the latest astronomical developments through courses and educational workshops, and guide visitors on inspiring journeys through the Universe in the Peter Harrison Planetarium.

This book contains over 100 puzzles, all of which have been inspired by the Royal Observatory's history and collections, including astronomical instruments, star charts, clocks, famous astronomers and their discoveries. You will also explore some of the latest astronomical theories and achievements in space exploration as your general knowledge and problem-solving skills are put to the test.

Try the puzzles in any order you like – some may be trickier than others, but you can always try again when you have picked up a bit more astronomical knowledge from elsewhere in the book. By the time you have solved them all, you will have been on a fascinating journey through space and time and unravelled some of the many mysteries of the Universe!

CHAPTER 1

TIME AND PLACE

Since the earliest times humans have looked to the skies to keep track of time and work out their location. In this chapter, put your knowledge to the test as you solve quizzes, number puzzles and anagrams about time and place. Discover how our knowledge has developed through the ages, from astrolabes to navigational charts and from sundials to Greenwich Mean Time.

NAVIGATIONAL TOOLS

Each of the clues on the next page describes a tool used by early navigators to find their position at sea, calculate the time, or keep track of astronomical findings. Work out what device each clue is describing, then enter it into the boxes opposite, one letter per box.

One letter is shared by all of the solutions, indicated by shining yellow squares. When completed, the letters in the starred squares will reveal, from top to bottom, the name of a lunar event that helped early astronomers to create detailed predictions of the positions of celestial bodies.

Left: *An instrument used for telling the time at night by measuring the position of the Great Bear or Little Bear constellations relative to the pole star. These devices are known to have been used since the tenth century, but this version was made in England in the 1600s.*

1. An ancient instrument used for calculating the height of celestial bodies above the horizon, prior to the invention of the device in clue 6

2. A device that uses magnetic forces to find north

3. Now commonly used to describe any annually published handbook, this record of statistical information comes from the Spanish-Arabic word for the device described in clue 11

4. A time-telling device that relies on shadows cast by the nearest star to Earth

5. This device consists of a large red sphere, which drops from its position at the Royal Observatory in Greenwich every day at 1 p.m. local time

6. A larger and more popular version of the device in clue 7, whose dimensions were determined by being one sixth of a circle

7. A device used for measuring the height of celestial objects, whose name derives from its shape, which is one eighth of a circle

8. A marine map

9. A primitive device thrown from ships to determine their speed at sea

10. A device used for measuring the time at night; this word is now most commonly used to describe animals that are active at night

11. A tool commonly used today to keep track of days, weeks and months

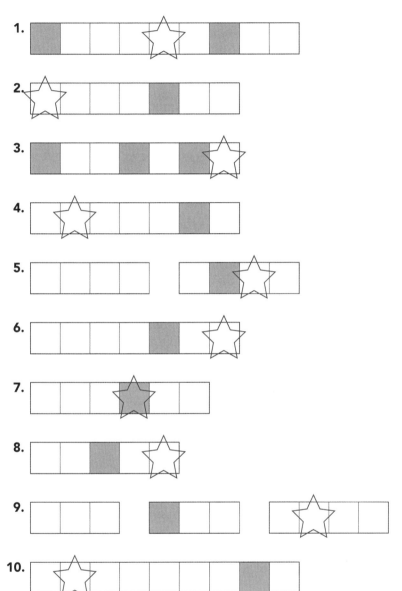

MULTITASKING MACHINES

The astrolabe is an astronomical instrument that was invented in ancient Greece and further developed by Islamic scholars during the Middle Ages. Before the development of modern technologies, astrolabes were incredibly useful astronomical calculators with many different applications. By lining up the different rotating parts of the instrument, astronomers could use astrolabes to calculate the time and date, the position of the Sun and stars, and to measure angles. They were also used to make astrological predictions and, by the fifteenth century, to aid in celestial navigation.

An astrolabe believed to date from the 1550s. It was made in the Belgian city of Leuven by a monk named Michael Piquer.

An astronomer has collected notes on a set of observations, hoping to use their findings to create a nautical chart. The astronomer has recorded details on three constellations during three different seasons, and also noted the level of the tide – either high or low – during each observation. Their notes are unclear, however, so it is up to you to try and work out what three combinations of night-sky constellation, season and tide were recorded.

Here is what you are able to discern:

- Leo was not observed at low tide
- Only notes on two constellations were taken at high tide
- Scorpius was seen in the dry season
- Cassiopeia was not observed during the rainy season
- The constellation viewed during harvest season was observed when the tide was in

Once you have made sense of it all, fill out this table of observations:

CONSTELLATION	SEASON (RAINY/DRY/HARVEST)	TIDE (HIGH/LOW)

MAP OF WINDS AND MAGNETISM

The map on pages 10–11 was made around 1715 and focuses on the Pacific Ocean. Arrows and finely engraved lines provide information about trade winds and wind directions at certain times of the year and isogonic lines at five-degree intervals detail magnetic variation, or the angle of variation between magnetic north, on a compass, and true north. The isogonic lines connect positions where the magnetic variation was the same at that point in time. Isogonic lines were a relatively new inclusion on charts at this time, having been first introduced by astronomer Edmond Halley 15 years earlier.

Use the excerpt from the map overleaf, and your general knowledge, to answer the following questions:

1. An inscription on the map notes a line that it calls the 'first meridian'. Which island does it say is used as a reference point?

2. Nearest to which marked 'line of variation' can you find a label that appears to describe two seemingly contradictory weather phenomena? The label is located in the middle of the Atlantic Ocean.

3. According to this map, how many degrees of variation from north would be shown on a compass in the 'Bermudas'?

4. Which line of longitude marked with a vertical line on the map corresponds most closely to the modern Greenwich meridian?

5. What degree of variation is shown by the line running through the area where modern-day Argentina is located?

6. What is the approximate difference in compass variation between the Azores and the island of Tristan da Cunha?

7. On which island shown on the map, long-since connected to the mainland with bridges, would a compass show approximately 8 degrees of west variation?

8. When spring begins in Brazil, what direction does the coastal wind blow, according to the map?

Overleaf: A section of a chart detailing wind directions, trade winds and magnetic variation between 50 degrees north and 50 degrees south of the equator, focusing on the Pacific Ocean. Made by London-based cartographer Herman Moll in around 1715.

PRIME MERIDIANS

Before Greenwich was chosen as the location of the international prime meridian in 1884, map-makers were at liberty to choose any line of longitude as a prime meridian for their map.

The map opposite has five historical meridians marked on it, each of which was frequently chosen as a prime meridian. The dots on each line mark the place each meridian is named after.

Can you use your geographical knowledge to guess the name given to each meridian, then write those names in the boxes below, with one letter per box. Each line has a clue to help you, and some letters are already given.

When complete, the letters marked in yellow will spell out the name of another common prime meridian, which continued to be used by its home nation until 1911 after it abstained from voting in the 1884 conference.

1. Island country

| C | | ▮ | | | | | | | | | E |

2. Group of islands

| C | | ▮ | | | | | | | | | | S |

3. Group of islands

| A | | ▮ | | S |

4. City borough

| G | | | | | ▮ | H |

5. Body of water

| B | | | | | | ▮ | | | | T |

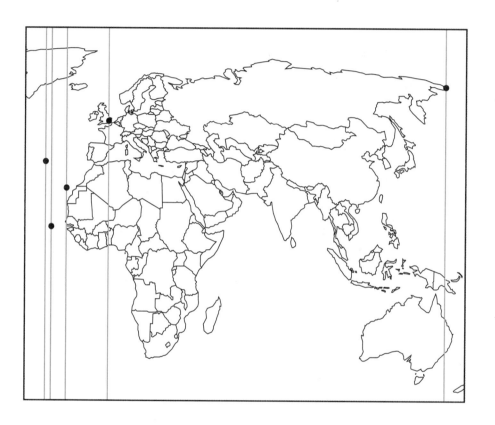

PRIME EXAMPLE

The prime meridian, marking longitude 0°, illuminated at night at the Royal Observatory in Greenwich. The meridian passes through a large, specially constructed telescope called the Airy Transit Circle, built in 1850 by the seventh Astronomer Royal, Sir George Biddell Airy.

Can you rearrange the order of the meridians below to reveal the names of eight countries or continents through which the Greenwich Prime Meridian passes? For example, you might insert meridian 5 before the first column, so that the first line starts with 'RN', the second with 'PI', and so on.

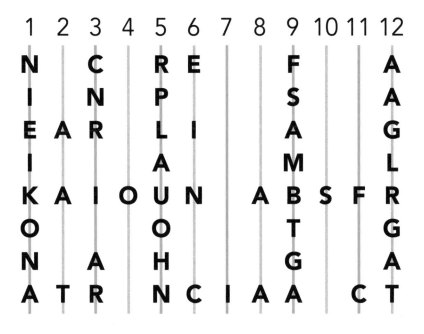

1	2	3	4	5	6	7	8	9	10	11	12
N		C		R	E			F			A
I		N		P				S			A
E	A	R		L	I			A			G
I				A				M			L
K	A	I	O	U	N		A	B	S	F	R
O				O				T			G
N		A		H				G			A
A	T	R		N	C	I	A	A		C	T

TIME AND TIDE

Rock, in Cornwall, at low tide, c.1906.

Ocean tides are produced by the gravitational pull of the Moon and the Sun upon the Earth. The frequency of tides depends on the passage of the Moon around the Earth – in most places around the world there are two tides each day, which vary in time by just under an hour a day. The height of the tide in a particular place depends on the coastline and nearby continental shelf, which is why the effects of the tides are more apparent on the coast than in the middle of the ocean. It is often not realised that air and landmasses also move due to tidal forces, sometimes experiencing a metre of vertical shift!

Sixteen different words have been jumbled up in the puzzle below. Eight of them describe **tides**, and eight describe **times or periods of time**. Each line contains both a time and a tide, with all of its letters in the correct order – but the two words have then had their letters mixed in among one another.

To help you, in each pair the 'time' word and the 'tide' word both have the same number of letters. How quickly can you find all 16 words?

MODIRUNRINANLG

NONOEANP

SMPIRNINUGTE

MSONLATCHK

DEABBY

NFLIGOOHDT

WHEIGEHK

NOLWOW

Once you have found all eight tide-related words, see if you can classify them as follows:

1. Join six of them into three pairs of 'opposites'.

2. Which word indicates a cycle of one high and one low tide in a day?

3. Which word describes a lack of tidal flow?

ISLAND TIME

A meridian is a line of longitude that divides Earth between east and west. Longitude is measured in degrees from the Prime Meridian in Greenwich. Because it is a distance in Earth's daily rotation, the difference in longitude between two places can also be thought of as the difference in their local times. A 15-degree change in longitude is equivalent to a time difference of one hour or, put another way, Earth turns through one degree of longitude every four minutes.

On this map of a fictional land, lines of latitude and longitude have been marked alongside a prime meridian. The bold lines of longitude are separated at 15-degree intervals, as they often are on global maps. Can you use both the map and your knowledge of time and longitude to answer the following questions?

1. What is the time difference between locations A and B?

2. At which labelled point do the clocks show a time 30 minutes ahead of the prime meridian?

3. How many degrees of longitude separate points C and G?

4. If the time at point G is 6.30 p.m., what is the time at point D?

5. If the time at the prime meridian is 4 a.m., at which location is it 8.30 a.m.?

6. A sailor is 75 degrees east of point A, whose local time is 3 p.m. What is his labelled location, and what is the local time at this labelled location?

7. A sailor calculates the time at his location to be 1.46 a.m. His clock, set at prime meridian time, shows 9.46 p.m. on the previous day. Which labelled location is he at?

8. A sailor travels from a port at point B to a port at point G. The journey takes ten hours, and his watch, set to the prime meridian time, shows 7.55 a.m. when he leaves.

- What is the local time at port B when he departs?

- What is the local time at port G when he arrives?

- Which labelled location shows a local time of 2.55 p.m. when he is halfway through his journey, according to the time travelled?

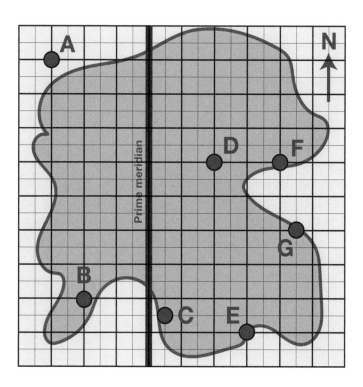

LOCAL TIME

An early nineteenth-century gold watch that displays both Greenwich Mean Time and another local time.

Before Greenwich Mean Time was used as the standard across the country, towns and cities in the UK kept their own local times. Local time could differ by as much as 20 minutes from London, creating many small time zones. In fact, a standard London time was only rolled out across the UK as a result of train travel, since minor time differences meant that schedules were difficult to coordinate, making accidents likely and trains far too easy to miss. Until this point, stations would often have one minute hand for local time and another, smaller hand for London time.

Try mapping out a range of small local time zones on the grid below, according to the following rules:

- Every empty square must contain a number (representing how many minutes ahead of London it is). Some squares already contain numbers, which cannot be changed.
- Every number must form part of a continuous region of as many squares as that number indicates – so a '3' must be part of a region of 3 squares. Regions are continuous wherever two squares of the same value touch, not counting diagonally.
- Two different regions made up of the same number of squares *cannot* touch, except diagonally.

You might find it helpful to draw borders between regions as you solve.

	2			2	1		2
				5		1	
	13	6		6	7		
		1			6		3
					2		
2	2		13	5			3
3					3		
			2		4		

A SUBSCRIPTION TO TIME

Maria Belville, who delivered Greenwich Mean Time to clockmakers in London using a pocket watch. When Maria retired aged 81 in 1892, her daughter Ruth continued the service until the 1930s.

Before the advent of the talking clock or modern digital services, clockmakers in London could sign up to a service to regularly receive the latest time from the Greenwich Observatory, so that they could update their timepieces accordingly. Initially the service was provided by the Belville family, members of which would travel around London in person with a pocket watch, updating horologists on the latest time. Eventually this service was replaced by an electrical system.

To test your ability to convey times, can you link up each pair of clocks showing the same time in this grid-based map of London below?

To do so, simply draw a series of separate paths, each connecting a pair of identical times. No more than one path can enter any square, and paths can only travel horizontally or vertically between squares.

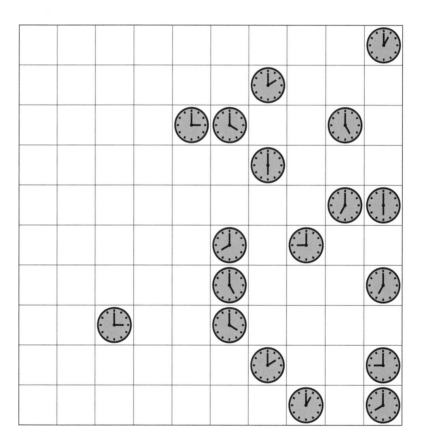

CRYPTIC CALENDARS

The names of four calendars have been cryptically concealed on the calendar below. Can you crack the code to reveal the calendar names?

M	T	W	T	F	S	S
		1 1 8	**2** 3	**3** 5 1	**4**	**5** 4 5 7 3 2 5
6	**7** 1 4	**8** 2 1	**9** 3 7	**10**	**11**	**12**
13	**14** 4 9	**15** 5	**16**	**17**	**18** 2 6 4	**19** 6
20 3	**21**	**22**	**23** 6	**24**	**25**	**26** 2
27	**28**	**29**	**30**	**31**		

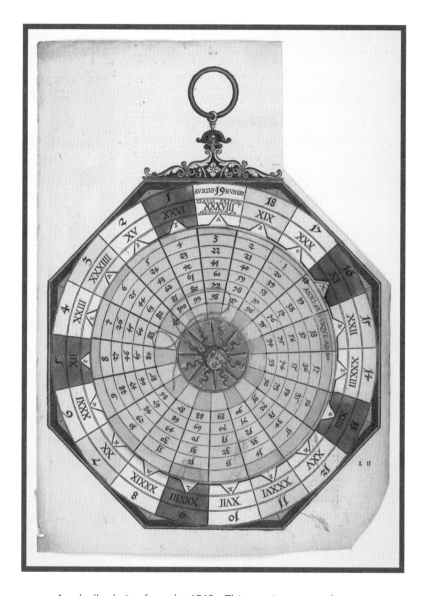

A volvelle dating from the 1540s. This rotating paper chart was used to find the quantity and dates of movable feasts necessary for the establishment of the Julian calendar.

ANNUAL ANAGRAMS

Hidden in the lines below are the names of six **periods of a year** that have each had their letters scrambled. They have also all had two extra letters added that spell out two further words.

The two new words are both **phenomena caused by the Earth's tilt**. Each line contributes one letter to each of these two words, although not necessarily in a consistent order, which can then be read in order from the top line to the bottom line.

Can you discover the six scrambled periods of the year, and the two phenomena that have been hidden among them?

WARY JANUS

CURIE EMBED

SUN GAMUT

RUBY FERMAT

EYEJUL

REJ RUN

Once you are complete, can you split the hidden words into two equal-sized sets?

PARTS OF A CLOCK

The movement of the first successful marine chronometer. By keeping the correct time at sea, it solved the problem of sailors not being able to determine their longitude, and therefore their accurate position east–west, at sea.

Can you unscramble these horology-themed words to reveal the name of a celebrated inventor?

The letters in the names of eight constituent parts of a clock have been scrambled. To make it even trickier, each clock part has also had one extra letter added to it. When these extra letters are extracted, and read in order from top to bottom, they will spell out the surname of the designer of the first successful marine chronometer.

Extra spaces have been added to the anagrams below, but each line is an anagrams of a one-word answer plus the one extra letter.

PLUMED HUN

CAROLLER NOT

AMPERE SCENT

PAL TER

ILL PAIR

BLANC SEA

ARM FOE

NOT VIP

WATCH THE CLOCK

The image shown below is a sketch made by John Harrison, showing part of the design for his fourth marine timekeeper, 'H4'. The resulting clock was the first successful marine chronometer, helping sailors to keep track of time and calculate their longitude.

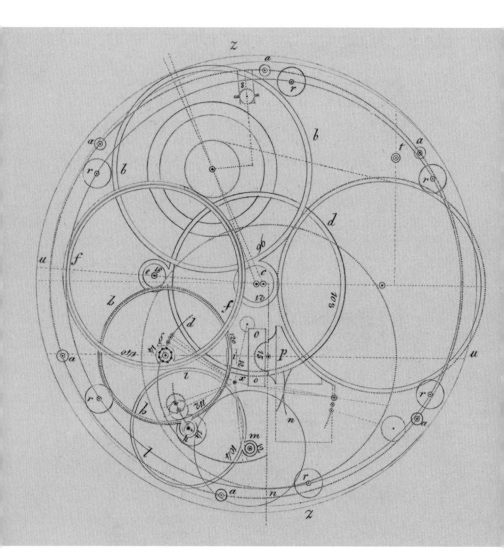

The original diagram is taken from *The Principles of Mr Harrison's Timekeeper*, 1767.

Can you spot five differences between the original sketch on the opposite page and the modified image below? The modified image has been rotated to make the task trickier.

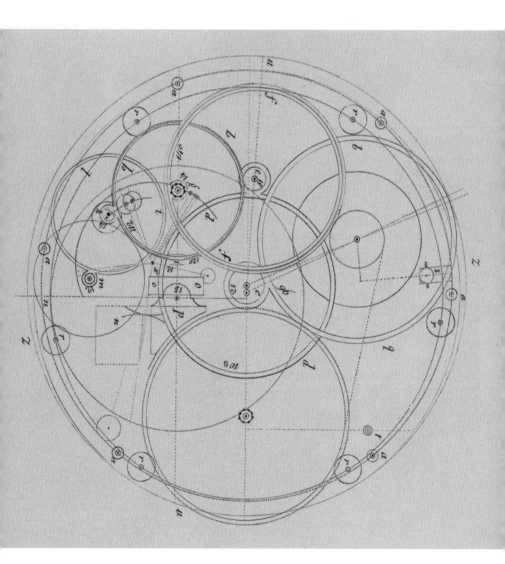

DIVISIONS OF TIME

A small portable sundial known as a butterfield dial, made in Paris in the late seventeenth century.

Can you fill in the clock dial opposite with the names of various units of time, both common and unusual?

Enter one letter in each box to spell out nine words, each of which fits its numbered clue given below. Yellow boxes are shared between words, so the last letter of one word becomes the first letter of the next.

When the clock is complete, letters with stars next to them will spell out – in the order of the numbering within the stars – the name of a medieval subdivision of time. One of these medieval units lasted the equivalent of 90 modern seconds, and the units were used in conjunction with sundials.

1. A day on Mars

2. Period of five years, taken from the name of an ancient Roman ritual that followed each census

3. Period of 365,242 days

4. Unit of time equal to sixty of the SI unit of time

5. A thousand million years, in astrology

6. A thousand-millionth of the SI unit of time

7. Period of 87,600 hours

8. Long period of history, of imprecise length

9. 'Year', in Latin

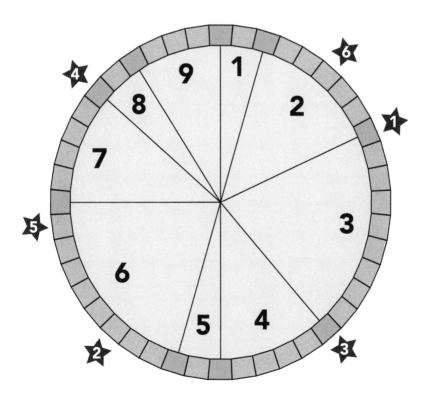

SHAKESPEARE'S TIME

Jumbled up across these pages are eight lines taken from plays by Shakespeare, all of which mention time. Restore the lines by linking one part from the first section, and a second part from the second section.

Better three hours too soon	and now doth time waste me
We are time's subjects,	and time bids be gone
I wasted time,	be master of his time
Let every man	for all things
There's a time	prologue
What's past is	plays many parts
Time and the hour runs	than a minute too late
And one man in his time	through the roughest day

Once you have restored the lines to their full length, can you match each line to the play, act and scene in which it is spoken? The lines are taken from the following scenes:

As You Like It, 2.7

Henry IV Part 2, 1.3

Macbeth, 1.3

Macbeth, 3.1

Richard II, 5.5

The Comedy of Errors, 2.2

The Merry Wives of Windsor, 2.2

The Tempest, 2.1

Emma, Lady Hamilton posing as the Shakespearean character Miranda. A print after a painting by George Romney, 1809.

TIME TEASERS

Can you decode these five well-known expressions featuring the word 'time', which have been shown as visual representations of their meanings?

1.

GOODALLTIME

2.

3.

4.

5.

LOST

7:48 AM

REWARD

CLOUD SPOTTING

Meteorology – forecasting the weather – became increasingly important to astronomers as they began to build more complex pictures of the night sky. Clouds blocked the view, and if you couldn't *see* the night sky, then you couldn't study it. It was in astronomers' interests to make use of meteorology as a predictive science.

Try using your logical deduction skills to make your own prediction of cloud locations in the puzzle below, shading in the locations of clouds in this grid-based map of the sky.

- Form clouds by shading square or rectangular areas of at least two squares wide and two squares tall.
- Clouds cannot touch, not even diagonally.
- Numbers outside the grid reveal the number of shaded squares in their row or column.

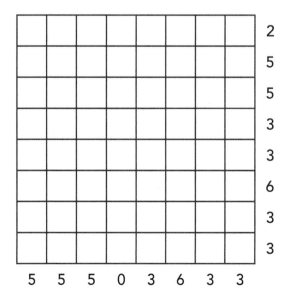

PEA-SOUPER LETTER SOUP

During the nineteenth century, the Royal Observatory became increasingly involved in studying changes in terrestrial magnetism and its effects on compass navigation. Changes in atmospheric pressure can have a significant effect on the instruments used by astronomers to measure activity in the Earth's magnetic field. Consequently, the Royal Observatory's meteorological observations were also expanded, and in 1838 a new Magnetic and Meteorological Department was established, reflecting the close relationship between the two relatively new areas of science. Knowing that a storm was brewing, for example, might help an astronomer to interpret unusual readings in the magnetic data.

Can you detect the storms brewing in the letter cloud below? The names of five different types of storm have been mixed up in the disturbance. Use every letter once to spell out and therefore restore these five words.

Once you have found the words, do you notice anything that these tempestuous words all have in common, beyond their meteorological meanings?

WAVES AND WEATHER

James Reynolds's astronomical print on the next page illustrates numerous meteorological phenomena, all of which were originally labelled on a caption beneath the image.

Use your knowledge of the weather to assign each of the events listed below to one of the numbered labels, 1–17, on the image.

Aerolites	Mock Suns
Aurora borealis	Perpetual Snow
Clouds: Cirrus	Rain
Falling Stars	Rainbow
Fog	Snow
Glaciers	The Maelstrom
Halo	Waterspouts
Lightning	Zodiacal Light
Mirage	

Once you have decided which label is which, try the following related questions:

1. Which one of these phenomena is also known by the more technical term 'parhelia'?

2. Which of the named phenomena derives from the Dutch name for a mythical whirlpool found off the coast of Norway?

3. Which of the phenomena may be caused by the reflection of sunlight on cosmic dust?

One of a set of 12 hand-coloured astronomical prints with an explanatory card, published by James Reynolds in the second half of the nineteenth century. Affordable prints like these were designed to appeal to middle-class Victorians who wanted to learn more about science at home. This diagram details various meteorological phenomena; the other prints in the set discussed topics including the seasons, the planets, and the phases of the moon.

MH
DISPLAYING THE

13
14
4
10
8
12

Drawn and Engraved by John Emslie.

Published by J. Reynolds, 174, Strand, Sep.r 20.t 1846.

Magnitudines Stellarum

CAROLO II DG MAG BRIT FRA HIB REGI SE
Hanc AUSTRALIS HEMISPHÆRII Ta
nuperis observationibus jussu Regio susceptis, r
plurimiis, stellis nondum Globo ascriptis, locu
submisse offert Subditus Humillimus
Edmon.Halleius. è Coll.Reg.Oxon.

Indus

Pavo

Grus

Piscis

Austrinus

Toucan

Fomalhaut

Aquarius

Phœnix

Achernar

Hydrus

Xiphias

Tropicus

Eridanus

Ca

Limen
Ptolium

Nova

Lepus

Regel

Cetus

Linea

Pol.Eclip.

Pol.Mundi

Qued.lol

Aries

CHAPTER 2

ASTRONOMICAL HISTORY

The mysteries of space have fascinated humans for thousands of years, and early civilisations each developed their own explanations of what they saw in the night sky. The long history of astronomy is the focus of this chapter. Solve the word puzzles, quizzes and codewords on the following pages to learn more about famous astronomers, the mapping of the stars and the discoveries that have led to the understanding of outer space we have today.

THE FIGURE OF ASTRONOMY

The figure of Astronomy, opposite, is accompanied with the ten letters that spell her name in Greek. Can you use these ten letters to solve the ten clues below, and reveal ten astronomical words or names? The number of letters in each solution is given, and letters may only be used per answer as many times as they appear on the sculpture.

1. The Sun, for example (4)

2. Charged particles (4)

3. US space agency, for short (4)

4. Latin word for the plural of answer 1 (above), used in the Brad Pitt movie title 'Ad ___' (5)

5. Units of matter once thought to be indivisible (5)

6. Word used in constellation names to mean 'lesser', as in 'Ursa ___' (5)

7. Natural planetary satellites (5)

8. Prominent constellation visible worldwide, with a well-known belt (5)

9. Mythology from which many planets in the Solar System derive their names (5)

10. Inhabitants of the fourth planet from the Sun, perhaps (8)

Astronomia, the Greek muse of astronomy, holding the Sun and the Moon in her hands and surrounded by signs of the zodiac. This sculpture was created by artist William J. Neatby in 1895 and decorates the outside of the South Building at the Royal Observatory in Greenwich.

ORDERLY ORRERIES

The object pictured here is an orrery, a model of the Solar
System invented in the early 1700s by renowned clockmakers
George Graham and Thomas Tompion. Orreries use gears to
make small models of the planets and moons rotate around the
Sun, demonstrating their relative positions and movements.
They became very popular in the eighteenth century, both as
educational tools and as curiosities. Orreries demonstrated the
heliocentric, or Sun-centred, model of the Solar System, which had
only become generally accepted in the previous two centuries.
Before that the geocentric model, which stated that the Sun and
planets rotated around the Earth, had dominated scientific thought
for more than 1,000 years.

The orrery shown opposite depicts several planets and their
moons. Based on the information presented here, what range of
years can the construction of the orrery most likely to be dated to?

First four moons of Jupiter discovered: 1610

First five moons of Saturn discovered: 1655–1684

Planet Uranus discovered: 1781

First two moons of Uranus discovered: 1787

Two more moons of Saturn discovered: 1789

Planet Neptune discovered: 1846

Neptune's first moon discovered: 1846

Eighth moon of Saturn discovered: 1848

Two more moons of Uranus discovered: 1851

A portable orrery made by scientific instrument maker William Jones (1763–1831).

THE LANGUAGE OF SPACE

As astronomers and explorers discovered more about outer space, they often came up with creative descriptions of what they were seeing. Can you match these eight astronomical terms to their etymology?

COMET

CRESCENT

ECLIPSE

GALAXY

GRAVITY

MOON

NEBULA

SOLSTICE

From Latin, meaning 'mist'

From Latin, meaning 'sun' plus 'stopped'

From Greek, meaning 'long-haired'

From Latin, meaning 'growing; to grow'

From Latin/Greek for 'month', and Latin for 'to measure'

From Greek, meaning 'milky'

From Latin, meaning 'seriousness; weight'

From Greek, meaning 'fail to appear'

EXTRA-TERRESTRIAL EXPLANATIONS

Can you match each of the astronomical terms below with one of the definitions beneath? Some decoy definitions – which don't apply to any of the words – have been added to make it trickier.

ALTAZIMUTH

ISOGONIC

NADIR

PARALLAX

OCCULTATION

SIDEREAL

ZENITH

A. A rocky structure found on high-velocity comets

B. A telescope mounting that allows movement about both a horizontal and vertical axis

C. Indicating points where variation in the Earth's magnetic field is of equal deviation

D. Official name for the dark side of the Moon

E. Relating to the distant stars

F. The apparent change in the position of a celestial body when viewed from different places

G. The movement of one celestial body in front of another, blocking the latter from view

H. The point on the celestial sphere that is directly above an observer

I. The point on the celestial sphere that is directly below an observer

READ SHIFT

This eighteenth-century astronomer became famous after discovering the planet Uranus and later also identified two of its moons. Solve the puzzle to discover his name.

The names of five famous astronomers have been encoded using a Caesar shift, whereby each letter has been shifted forwards by a certain number of positions in the alphabet. The word ABACUS with a shift of 2, for example, would have become CDCEWU, where A became C, B became D and so on, until Y became A, and Z became B.

The names have all been encoded with different sizes of shift. The clues list one achievement of each astronomer, and the number of items listed in the clue is the shift number that was used to encode that astronomer's name. The coded names and achievements are not in the same order, however, so it's up to you to work out which achievement – and shift – belongs to which encoded name.

OJDPMBVT DPQFSOJDVT

MRKDQQHV NHSOHU

PZHHJ ULDAVU

KMSZERRM GEWWMRM

YKNNKCO JGTUEJGN

Discovered **four** moons of Saturn

Discovered **two** moons of Uranus

Demonstrated that light could be split into the **seven**
colours of the rainbow

Proposed that planets move around **one** body – the Sun

Proposed **three** laws governing planetary orbital motion

SPIDER WEB

When Charles II established the Royal Observatory in 1675 he appointed John Flamsteed as the first Astronomer Royal. Flamsteed was the first person to live and work at the Observatory in Greenwich and remained in the position until his death in 1719. The king tasked him with the job of mapping the night sky and producing data accurate enough to aid with navigation at sea. One of Flamsteed's achievements was his production of the first detailed star map of the northern hemisphere using a telescope. At the time astronomers often used spider's silk – a strong, fine material – to create crosshairs on their lenses, allowing for more precise measurements to be taken.

John Flamsteed (1646–1719), the first Astronomer Royal. Based on a portrait by Thomas Gibson and engraved by George Vertue in 1721.

Can you find the name Flamsteed in the spider web below? Start on any circle and then follow lines to touching circles, so that each circle visited in turn spells out the word. No circle can be revisited.

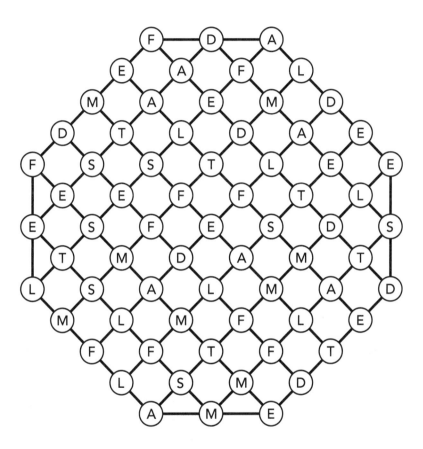

A REGULAR VISITOR

Passing comets are often visible from Earth with the naked eye, and as such they have fascinated humans for centuries, although it is only in more recent history that scientists have begun to properly understand them. In the early 1700s, the astronomer Edmond Halley studied 24 comets that had been seen from Earth over the previous 400 years. After noticing that three of the comets had very similar characteristics, Halley deduced that they were actually the same comet that had passed by Earth several times. Further calculations led him to predict that the comet would next return in 1758. Halley died before then, but he was proved correct.

Halley's Comet, as it became known, was the first recognised periodic comet, meaning it has a well-known orbit that can be confidently predicted. Today many periodic comets have been identified, but Halley's Comet is perhaps still the most famous. It will next be seen from Earth in July 2061.

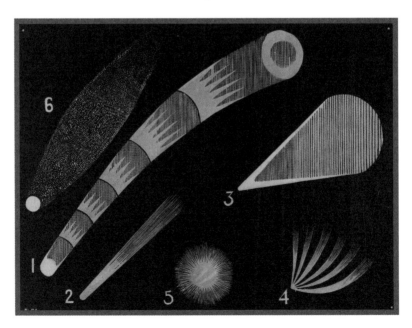

All of the following words or names have a regular appearance of one of the letters in 'HALLEY'. Can you fill in the blanks to reveal the original entries?

The definitions of the words and names are given below, but they are not given in the same order as the list below – so it's up to you to work out which definition clues which incomplete word or name.

H _ H _

A _ A _ A _ A

_ _ _ L _ _ _ L

L _ _ _ L _ _ _

_ E _ E _ E

_ Y _ Y _ Y

The alignment of celestial bodies in straight lines, as e.g., the Sun, Earth and Moon

Chilean desert where the Very Large Telescope is found

Rare word meaning 'brightly shining'

Greek goddess of the Moon

Occurring at regular intervals

Surname of the 'father of nuclear chemistry'

Left: A wall hanging featuring six comets seen from Earth since the 1600s. It was produced for a charitable society called the Working Men's Educational Union in the 1850s and would have been used in educational lectures.

STARRY SKIES

Since ancient times, humans have grouped nearby stars into constellations, which work rather like crude celestial dot-to-dot puzzles. The actual stars within a constellation may be nowhere near one another in reality, but from the point of view of the Earth they appear in nearby parts of the sky.

Constellations have been named after many different things, and through history have been used for a variety of astronomical purposes.

Can you work out what two classifications have been used to sort the six constellations shown in the Venn diagram below? Each star represents a different rule, and the area in the middle applies to both stars.

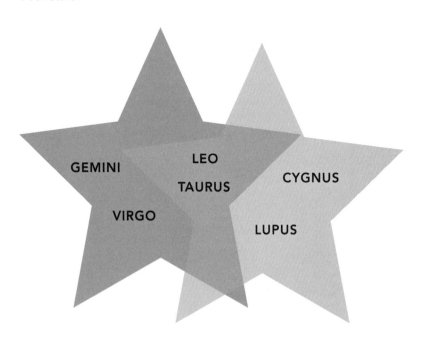

GEMINI

LEO

TAURUS

CYGNUS

VIRGO

LUPUS

OVERLAPPING WORLDS

In a similar way to the previous puzzle, can you work out what two classifications have been used to sort the six Solar System bodies shown in the Venn diagram below?

Each circle represents a different rule, and the area in the middle applies to both circles.

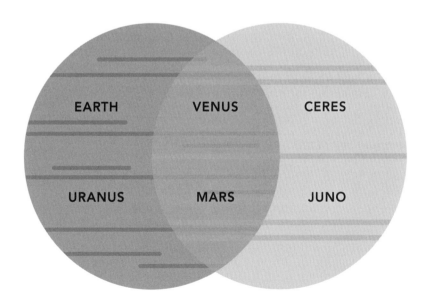

AN EPITOME OF ASTRONOMY

Can you use the historical chart shown on pages 58–59 to answer the following questions? Use only the figures given in the chart, even if you know that the numbers have since been superseded by more accurate measurements.

1. This chart shows seven planets. Which major planet – named after the Roman god of the sea – remained to be discovered at the time this chart was drawn?

2. Which British-German astronomer do you think was credited with the discovery of Uranus, given the name shown on this chart?

3. In what way does the visual presentation of the Solar System information reverse reality?

4. What do you think the unit of distance specified in the outermost ring of the diagram suggests about how distances were standardised at the time of the chart?

5. Which of these two planet pairs has the greater difference in diameter: Mercury versus Venus, or Earth versus Mars?

6. The chart shows estimates of the circumference of the various Solar System bodies.

- How many times larger is the circumference of the Sun compared to the circumference of the next largest Solar System body on this chart, to the nearest whole number?

- Which two planets' circumferences are shown as almost exactly one hundredth of the Sun's circumference?

7. A solar day is defined as the time taken for the Sun to return to the same place in the sky, and on Earth it is on average 24 hours long. This contrasts with a sidereal day, which is the time taken for a complete revolution about a planet's axis of rotation. On Earth this is 23 hours and 56 minutes, but on other planets it can be hugely different to the length of its solar day.

- Assuming that the chart shows the length of solar days, is a Venus solar day shorter or longer than an Earth day?
- According to modern science, do you know if a Venus solar day is actually shorter or longer than an Earth day?

8. To the nearest whole number, how many Earth years does a year on Jupiter last?

9. Which two planets can come closer to one another than any other pair of planets?

10. The speeds of the planets are given in the rather unusual unit of multiples of the speed of a cannonball. The actual calculations on the chart are inconsistent, but if you use just the values for Jupiter then can you say what the actual speed of a cannonball is being taken to be, to the nearest mile per hour? For the segment 'The degree of velocity in a minute in English miles', read this as 'Velocity in miles per minute'.

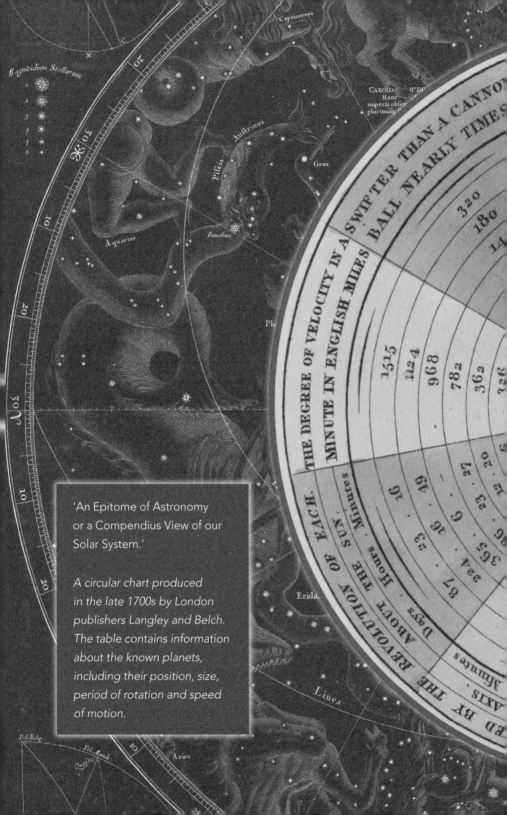

'An Epitome of Astronomy
or a Compendius View of our
Solar System.'

A circular chart produced
in the late 1700s by London
publishers Langley and Belch.
The table contains information
about the known planets,
including their position, size,
period of rotation and speed
of motion.

THE NAMES OF THE PLANETS.

THE NAMES OF THE PLANETS.	THE DIAMETER OF EACH.	ADMEASUREMENT BY ENGLISH MILES THE CIRCUMFERENCE OF EACH.	THE DISTANCE OF THE SUN FROM	THE ORBIT OR CIRCLE IT DESCRIBES.	TIME OCC... ON ITS OW... Days . Hou...
SUN	796.000	2.501.964			Uncertain
MERCURY	2.460	7.724	32.000.000	201,024,000	30.295 Days
VENUS	7.906	24.825	59.000.000	370,636,000	Uncertain
EARTH	7.964	25.020	81.000.000	508,939,200	Uncertain
MARS	4.440	13.960	130.000.000	773,686,000	24 . 40
JUPITER	81.155	254.908	424.000.000	2,662,280,000	9 . 56
SATURN	67.670	213.112	777.000.000	4,881,891,000	10 . 16
HERSCHELL	35.112	114.912	1,800,000,000		Uncertain

CONSTELLATION COMPUTATION

Humans have attempted to interpret and map the night sky for thousands of years. Ancient civilisations first grouped stars together into constellations as a way of establishing patterns in the sky and used them as early seasonal calendars. The two disciplines of astronomy and astrology were both born out of the desire to understand the stars – astronomy based in the search for knowledge about celestial objects, and astrology in the attempt to relate that knowledge to earthly events.

For many centuries astronomy and astrology remained closely linked together. In the royal courts of medieval Europe, for example, astronomers would often be called upon to make astrological predictions for the monarchs they served, regarding anything from illness to an auspicious date for a battle. It was only in the seventeenth century, as new scientific ideas and astronomical theories became more prominent, that the two disciplines began to diverge.

The names of four major constellations have been encoded in a particular way below. Can you work out the method that has been used to disguise them, and therefore reveal the names of the four groups of stars?

Illustrations of the four constellations are also given, opposite. Can you match each decoded constellation to its corresponding illustration?

UDCA	REEN	SLND
APTR	MHAO	AIUM
JNRE	OUUD	RSSA

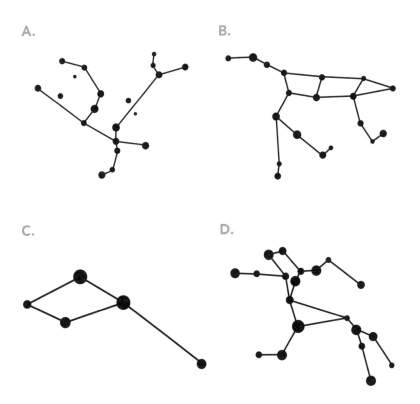

Once you have completed the matching, can you say which of these constellations shares its name with the galaxy that is predicted to collide with our own Milky Way in about four billion years?

ILLUSTRATED HEAVENS 1

The images of constellations on the following pages are taken from a set of nineteenth-century educational cards called *Urania's Mirror*. The set contains 32 constellation cards and a chart of the night sky for the latitude of London, all contained in a box decorated with the figure of Urania, the personification of Astronomy. Each of the hand-painted cards depicts the stars and popular imagery of a well-known constellation. The set was produced in 1825 and would have been sold as an educational tool for families to use at home, reflecting the popularity of improving pastimes like amateur astronomy in the late eighteenth and early nineteenth centuries.

How is your knowledge of the zodiacal constellations? Can you match each picture to the sign it represents? As an aide memoir, if needed, the names of the twelve signs of the zodiac are given below – but all of the letters in the word 'ZODIAC' have been removed:

GEMN   LBR   LE   NER

PRRN   PSES   QURUS   RES

SGTTRUS   SRP   TURUS   VRG

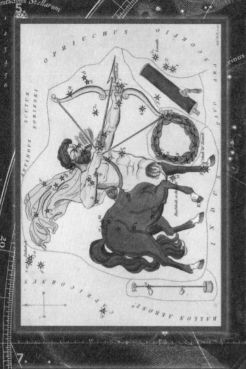

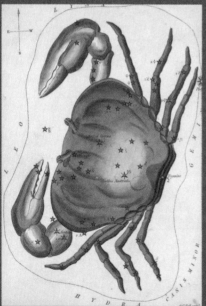

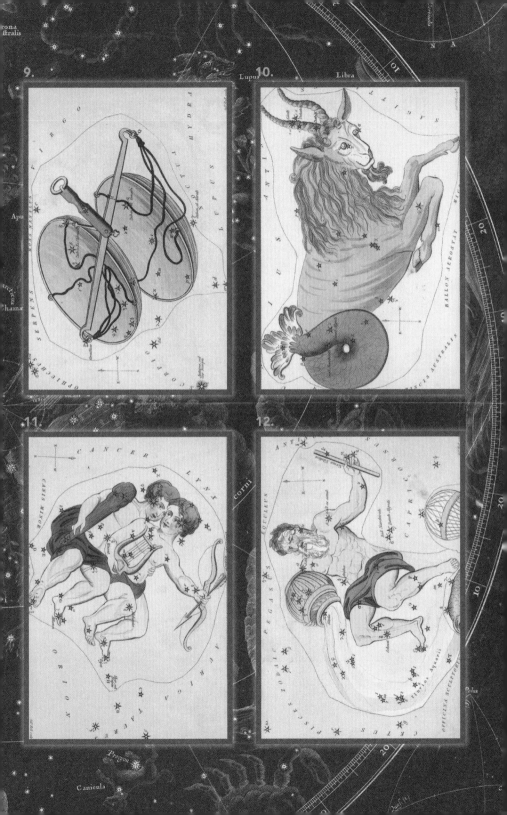

ILLUSTRATED HEAVENS 2

Referring back to the illustrations of the constellations in the previous puzzle, can you match each of the 12 illustrated constellations with one of the star patterns on these pages?

Connecting lines between stars have not been drawn in, so that the dots appear as they would in the night sky. How many can you pair up? Note that some of the star patterns may be rotated relative to their illustrated images.

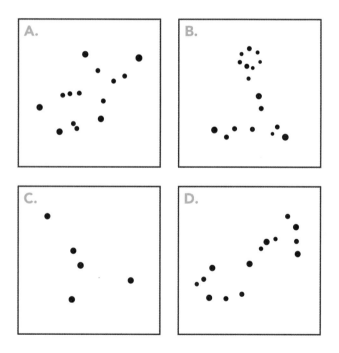

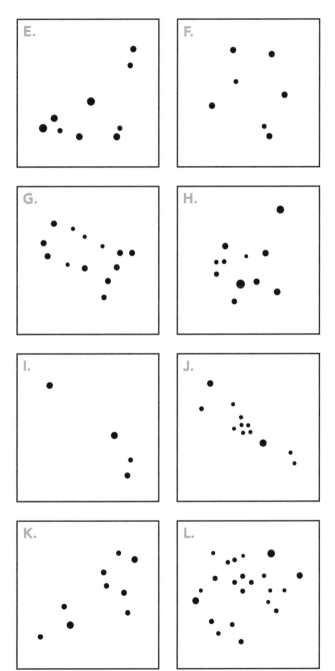

CONSTELLATION CHARTS

The image on the opposite page comes from the same pack of educational cards, known as 'Urania's Mirror', as used in the previous two puzzles. Can you use this star chart, and your knowledge of the skies, to answer these more in-depth questions?

1. Which constellation, partly shown on the star chart opposite and labelled with its Latin name, is perhaps most recognisable for its seven brightest stars, which form the 'Plough'?

2. What does the Latin name of the constellation referenced in the previous question translate to in English?

3. What is the gold object shown directly above the male figure's head? Its Latin name can be found in the bottom of the image.

4. The object in the previous question can measure a certain range of angles. What is the name of the similarly named tool used in surveying and navigation that can measure two-thirds of the range of angles that the previous object can measure?

5. Can you find a constellation on the star chart opposite with a Latin name meaning 'Northern Crown'? It may help you to recall the Latin name for an atmospheric phenomenon sometimes seen in the far north.

6. Can you find the name of a constellation that is now considered obsolete, and which depicted a Greek mountain? It includes the Latin word for 'mountain' in its name.

7. The fourth-brightest star in the night sky is shown on the star chart opposite. Its illustrated depiction gives a clue to its brightness, and its name also contains the letters of the word 'star'. What Greek letter is it marked with?

BOOTES, CANES VENATICI, COMA BERENICES, AND P.O.

QUADRANS MURALIS

SOUTHERN SKIES

On pages 72–3 is a printed star chart of the southern celestial hemisphere produced by the astronomer Edmond Halley in the 1670s. Halley travelled to the island of Saint Helena to catalogue the stars visible in the southern sky, and his observations helped establish his reputation as an astronomer.

The southern pole star, Sigma Octantis – labelled on this map as 'Polus Antarcticus' – is barely visible to the naked eye even in perfect conditions, unlike the prominent northern pole star, Polaris. The label for Polus Antarcticus is located on the central vertical line of the map, written in a tight curve, just up from the centre of the map, and the star itself is indeed barely visible on the map.

Can you use your map-reading skills and knowledge of the skies to answer the following questions on Halley's chart, which can be found on the following two pages? The writing is very small in places, so you may find it useful to use the magnification feature on a phone, or indeed an actual magnifying glass.

1. There is a key to the brightness of the included stars provided at the top left of the chart. Only one star on the chart is at the maximum brightness of '1'. What is its name, and what constellation is it in?

2. With the southern pole star essentially invisible to the naked eye, various other star-based astronomical methods are used to find which way is south. In particular the constellation labelled on this map as '*Crofiers*' contains two pointer stars that are often used for this purpose.

- This constellation appears on the flags of several countries in the southern hemisphere, but what is its English name?

- Can you name the five countries on whose flag the constellation appears?

3. Various birds are illustrated on the map, but which labelled constellation representing a tropical bird is depicted as closest to the southern pole star?

4. There is a constellation that is labelled in Latin with a name meaning 'Southern Crown'. It can be found in a line directly above the southern pole star. What is its name on the map?

5. Can you find the labelled celestial line of latitude that passes close to the three stars making up Orion's 'belt'? The label for this line of latitude is given in Latin (as two words, written with a large space between them), but what do you think is its modern English translation?

6. The word 'crater' is used in modern astronomy to describe the impact sites of asteroids on the Moon and other celestial bodies. Can you guess from the illustrations on the chart the approximate original meaning of this word, from which its modern meaning was derived?

7. Which prominent ship from Greek mythology is pictured on the chart, flying an English flag?

8. The circular area of the map is divided into segments which are indexed by the scale on the perimeter, with each segment covering 30 degrees of longitude. Each segment is also labelled, at the start of its scale, with a symbol (although some are cut off at the top or bottom of the page). You might recognise some of these symbols, so what do you think these symbols represent?

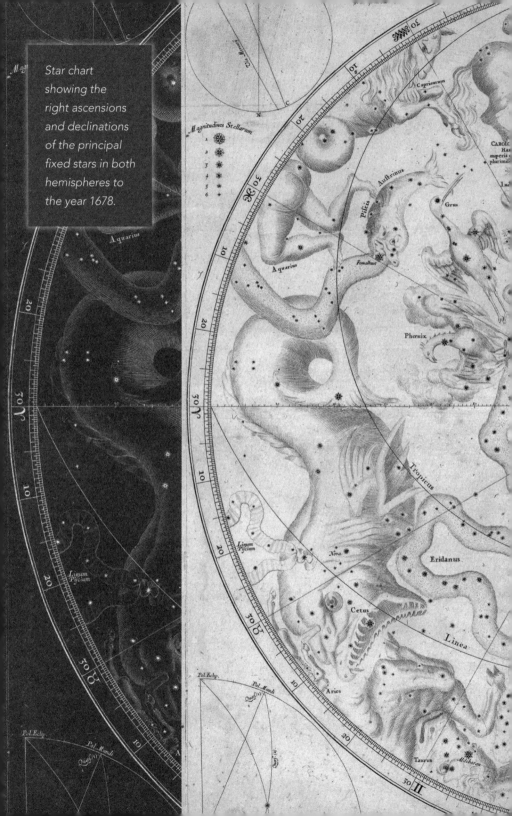

Star chart showing the right ascensions and declinations of the principal fixed stars in both hemispheres to the year 1678.

ASTERISMS

Traditionally, constellations have been thought of as groups of stars that form a pattern when viewed from Earth. The constellations we recognise today are still based on those named by the ancient Greeks, but in the 1920s an international meeting of astronomers formally defined 88 modern constellations and, for the first time, gave them borders. That means that today the term constellation refers not to a single pattern of stars, but to an entire region of the night sky.

Within any constellation there can be many smaller, informally identified groups of stars known as asterisms. For example, Orion's Belt is just one asterism within the constellation of Orion. The stars that make up an asterism can lie within a single constellation, or sometimes fall across several different ones.

The names of the six asterisms shown on these pages also only have part of their original names. They have been manipulated by removing any letters in their name which also appear in the word 'STAR', although any spaces between words have been preserved.

Can you use the pictures of the asterism star patterns, and the surviving letters, to work out what their full names should be? For example, the image below is labelled 'HE PLOUGH', since it has lost its 'T', and so can be restored to read 'THE PLOUGH' – which is itself one of the most recognisable asterisms in the night sky.

HE PLOUGH

1. PEZIUM

2. PING INGLE

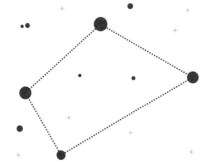

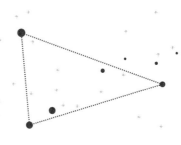

3. EVEN IE

4. EPO

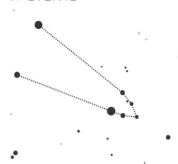

5. OION' BEL

6. GE QUE OF PEGU

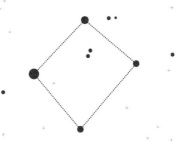

CELESTIAL DEITIES

The names of the following ancient mythological deities have had certain letters removed from their names, but these missing letters can all be found in the name of the single celestial body they are *all* associated with.

Can you restore the missing letters and reveal their astronomical connection?

SELEE – Greek

DIAA – Roman

ÁI – Norse

PHEBE – Greek

KHSU – Egyptian

CHAG'E – Chinese

ARTEIS – Greek

Once you have restored the deities, can you say which one shares its name with its country's space programme?

*A decorative allegorical panel by Sir James Thornhill,
depicting an assembly of gods and goddesses on Mount Olympus.
The message of the panel centres on the contest between reason,
personified by Apollo, the god of light and learning, and the passions,
symbolised by Cupid and Venus.*

SCRAMBLED SENSE

A nineteenth-century oil painting of Renaissance astronomer and mathematician Galileo Galilei (1564–1642).

One word in each of the following astronomical sentences has been changed so that its letters are scrambled. Can you work out which word is scrambled, and then unscramble the letters to restore the original sentence?

A small unable can be seen in the 'sword' of the Orion constellation

Copernicus has a lunar tracer named after him

Galileo was the first to divorces moons other than our own

Earth and Mars have ice caps on their slope

Scalloping stars sometimes form black holes

A lunar eclipse won't be visible from all ungodliest

The scattering of hustling through atmospheric particles makes the sky appear blue

Venus and Mars were once mistakenly considered wardening stars

ZODIAC ZIG ZAGS

Can you travel down from the word 'ZODIAC', at the top of the chain, to the bottom, by using your knowledge of scientific terms? Fill in the gaps with one letter per underline, so that an astronomical word matching its clue appears on each line. The two letters at the end of a word should be identical (and in the same order) to the two letters at the beginning of the next word, as shown by the lines.

Z O D I A C _ _   Pertaining to the twelve signs of the Zodiac

_ _ K A L I _ _   Opposite of acidic

_ _ U T R I _ _   Neutral subatomic particle

_ _ O N T I _ _   Literary term for midday

_ _ R I D I _ _   A line of longitude

_ _ A L E M _ _   Scale shaped like an '8' mapping the Sun

_ _ G N E T _ _   Type of field surrounding the Earth

_ _ E C A P S   Polar features on Earth and Mars

Magnitudines Stellarum

1
2
3
4
5
6

C Capricornus

N

20

☉

30

10

Piscis Austrinus

CAROLO II° D.G. MAG:BRIT: FRA; HIB; REGI S
Hanc AUSTRALIS HEMISPHÆRII T.
nuperis observationibus jussu Regio susceptis,
plurimisq; stellis nondum Globo ascriptis, loc.
submissè offert Subditus humillim.
Edmon.Halleius. è Coll: Reg: Oxo.

Indus
Pavo

Grus

Aquarius
Fomalhaut
Toucan

20

Phœnix

Acarnar
Hydrus

Xiphias

30 ♈

Tropicus

10

Linum Piscium

20

Nova

Eridanus
Ca.

Lepus

30 ♉

Cetus

Regul.

Linea

Pol.Echp.

10

Pol.Mund.

Quad.(?)

Aries

20

CHAPTER 3

THE SOLAR SYSTEM

Today we know that Earth is one of eight planets in the Solar System that all orbit our closest star, the Sun. This understanding of our corner of space is relatively recent, however; in fact, astronomers only began to call it the Solar System in the early 1700s. Since then, our knowledge has dramatically increased and we know more than ever before about Earth, its neighbouring planets, and even other planetary systems outside our own. Test your astronomical knowledge by solving the following puzzles about our incredible Solar System, and perhaps learn more about it along the way.

ATMOSPHERIC WORDS

Earth's atmosphere is a protective barrier that shields life on Earth from the Sun's heat and radiation. It contains the different gases that make up the air we breathe. The atmosphere is divided into five different layers that are defined by temperature and distance from Earth. The layer closest to Earth is called the troposphere – this area contains most of the water vapour in the entire atmosphere, which is why clouds are formed there. Next is the stratosphere, where the ozone layer shields Earth from the effects of the Sun's UV radiation. Above the stratosphere is the mesosphere, where meteors burn up as they approach Earth, and then the thermosphere. At this level, the atmosphere is very thin, and in the final layer, the exosphere, it blends with outer space.

The diagram opposite shows Earth and its various atmospheric layers, with the ending 'sphere' deleted from the name of each layer. Use the letters in the diagram to solve each of the following clues by spelling out each solution from left to right. For each word, begin at Earth and then and take **one** letter from each set of letters.

As an example, the word 'hot' would take an 'H' from EARTH, an 'O' from TROPO and then a 'T' from 'STRATO'. Note that letters *can* be reused between different clues.

For the first three clues, use only from Earth to the mesosphere, to spell four-letter words:

1. It typically consists of a nucleus and orbiting electrons

2. An asteroid that travels closer to the Earth than any other celestial body except the Moon

3. Major divisions of time

For these next three clues, use from Earth to the thermosphere, to reveal three five-letter words:

4. Animal that the constellation *Equuleus* is named after

5. Violent circular wave of air formed in the lee of a mountain or hill; also the name of a machine part

6. A geological formation consisting of a ridge of land that has been forced up between two fault lines

For the final clue, use from Earth to the exosphere to reveal a six-letter word:

7. One who might confront Taurus?

Once you have answered the above, can you say in which one of these layers that the International Space Station would be found?

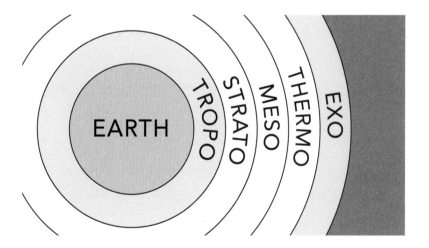

CONCEALED PLANETS

Can you find the names of six planets hidden in the sentences below? As an example, although not a planet, the word 'Sun' is hidden in the phrase 'I was **un**der a rain cloud'.

1. These clouds are probably going to mar some of the lunar eclipse viewings.

2. There are eleven usable emergency exits on the spacecraft.

3. The ground control bureau ran us through the list of spacewalk protocols before take-off.

4. Make sure everybody gets a turn using the telescope.

5. The landing made us look inept; uneven lunar surfaces are not ideal touchdown sites.

6. Can you actually hear the sound of the aurora borealis?

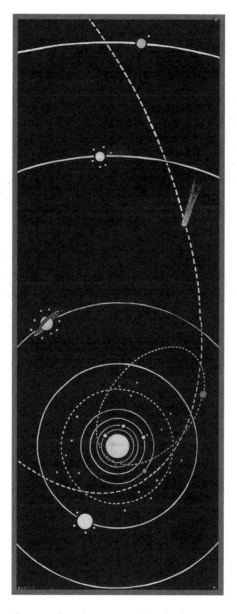

A wall hanging showing the planetary orbits, the known moons and other celestial objects including comets and asteroids. It was produced as a lecture illustration tool for the Working Men's Education Union in the 1850s.

GOLDILOCKS AND THE THREE THERMAL RANGES

This view of Earth rising over the Moon's horizon was taken from the Apollo 11 spacecraft.

Planet Earth is found in what is known as the 'Goldilocks Zone' of a solar system – an area that is neither too hot nor too cold to sustain life. In fact, the conditions needed to support life on Earth are 'just right'. Based on what we know about living things so far, planets inside 'Goldilocks Zones' in other solar systems also have the potential to sustain life.

Can you create a delicate balance of heat conditions in the puzzle below? Each coloured dot represents a different heat condition: red is too hot, blue is too cold, and green is just right.

Place a red, green and blue dot once each into every row and column within the grid, so that exactly one empty square remains in every row and column. Each coloured dot given outside the grid must match the closest colour to it within the same row/column. No more than one coloured dot may be placed in any square.

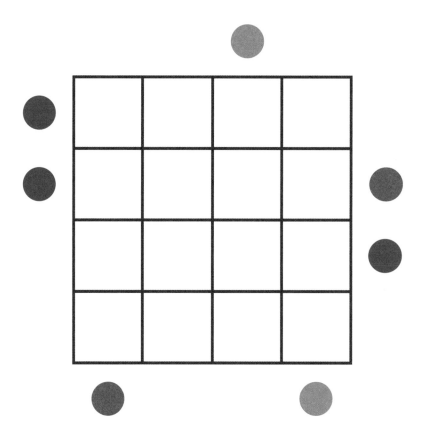

SOME SUN SUMS

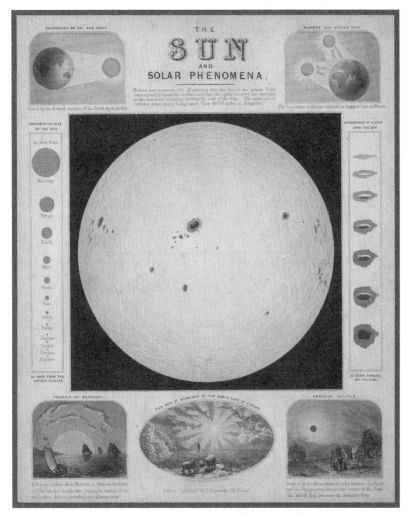

'The Sun and solar phenomena.' *One of a set of 12 hand-coloured astronomical prints designed to encourage learning at home, published by James Reynolds in the second half of the nineteenth century.*

Can you match each of the measurements labelled A–G with its corresponding value, below?

91   **8**   **1.39**   **330,000**

4.6   **15**   **13 billion**

A. Age of the Sun in billions of years

B. Approximate number of minutes it takes for the Sun's light to reach us

C. Diameter of the Sun, in millions of kilometres

D. How much brighter the Sun appears to us than the next-brightest star, Sirius

E. How many times greater the Sun's mass is than Earth's

F. Percentage of the Sun that is composed of hydrogen

G. Temperature of the Sun's core in millions of degrees Celsius

SOLAR ECLIPSES

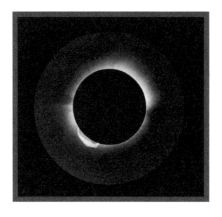

A glass photographic plate showing the eclipse of the Sun that occurred on 29 May 1919, as seen from Sobral, Brazil.

The diagrams on the opposite page each show a different configuration of the Earth, Moon and Sun, in various states of eclipse. They are not drawn to scale, but use their relative sizes and positions as shown for the purpose of answering the questions in this puzzle.

Can you work out approximately what each of these four configurations would look like, if viewed by an observer on Earth standing at the point marked with the red dot? Match each configuration to one of the four options below:

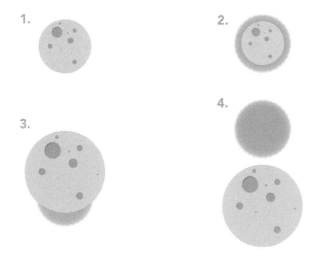

A.

B.

C.

D.

ECLIPSED DICTIONARY

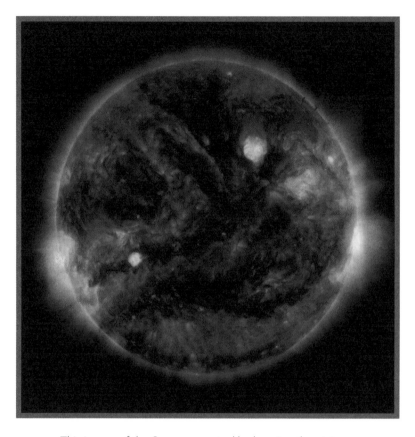

This image of the Sun was created by layering the pictures
produced by several different telescopes in April 2015.
The resulting image shows several solar microflares – explosions
of energy – erupting from the Sun's surface.

These eight Sun-related words and terms have all undergone a 'solar eclipse' of their own – that is, the letters of the word 'SUN' have been hidden (removed) from each of them. Can you restore the missing letters, then match each word with a definition below? Spaces between words have been kept in their original places.

CLEAR FIO

COROA

LMIACE

MADER MIIMM

OLAR WID

PHOTOYTHEI

POT

YODIC DAY

Historical period of time noted for its low sunspot activity, named after a famous astronomer

Dark patch on the Sun's surface

Measure of an object's brightness, with SI unit candela

Outermost edge of the Sun's atmosphere, visible during a solar eclipse

Stream of charged particles from the Sun, causing auroras

The process via which the Sun releases heat and light

The process that supports life using the Sun's rays

The time taken for a planet to rotate around its axis once in relation to its nearest star

ECLIPSE COLLECTION

A glass lantern slide composed of seven images taken in November 1910 showing the stages of a lunar eclipse.

Seven words used in describing astronomical eclipses have each been eclipsed by celestial objects on the opposite page, and in each case part of each word has been obscured.

Each of the objects is marked with the same word on both sides, and the seven objects are seen from both their fronts and their backs in the pictures opposite – although not in the same order.

Can you match the first half of each word (in the row on the left) with its corresponding second half (in the row on the right), to reveal seven astronomical terms?

Once you have restored the words, can you say which one is also used, in addition to its eclipse-related usage, to describe the darkest portion of a sunspot?

ANN LUN PAR PEN SO TOT UM

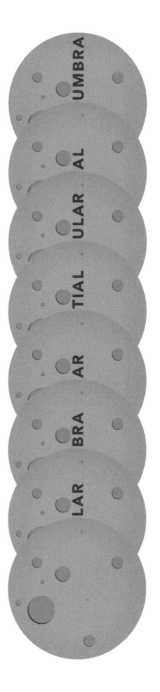

LAR BRA AR TIAL ULAR AL UMBRA

SOLAR DEITIES

Can you fit the names of these 12 deities, all of whom are associated with the Sun, into the grid opposite? One name should be entered on each line, with one letter per box, on the same row as the mythology with which it is associated.

When complete, the letters in the highlighted boxes can be rearranged to spell the name of an additional Sun deity. Can you work out who it is, and say which mythology they come from?

The names of the deities to fit into the grid are:

AINE	EKHI	HELIOS
INTI	MERI	RA
SAULE	SOL	SUÉ
SULIS	SURYA	XIHE

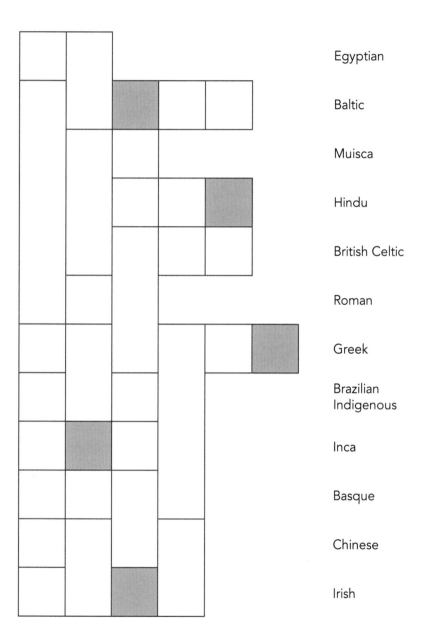

Egyptian

Baltic

Muisca

Hindu

British Celtic

Roman

Greek

Brazilian
Indigenous

Inca

Basque

Chinese

Irish

SUN SCAPE

First, solve the following clues so that each one results in a one-word solution. All eight of these solutions can be made into astronomical words by adding the word 'sun' to the start:

Start a phone call

Rupture

Front of a ship

Increase, like a price

Not heavy?

Notice (something)

Rafter

Soft feathers

Once you have found the eight words, enter each solution into one of the rows of boxes opposite, one letter per box, so that squares with identical colours contain the same letter. The clues are not given in the same order, however, so it is up to you to work out which solution goes where.

S U N

S U N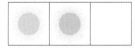

S U N

S U N

S U N

S U N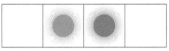

S U N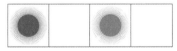

S U N

A SPOT OF SUN

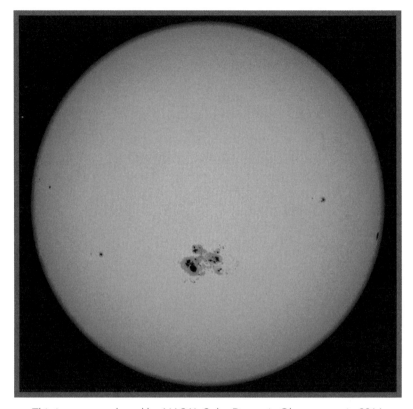

This image, produced by NASA's Solar Dynamic Observatory in 2014, shows an enormous sunspot on the Sun's lower centre. The sunspot, or dark region on the Sun's surface, is almost 80,000 miles in diameter.

Scientists are continuously working to understand more about the Sun, our closest star, and its effects on Earth. Using new technologies like NASA's Solar Dynamics Observatory, scientists are able to study the Sun like never before. The small spacecraft observes the Sun and sends data back to Earth, like this image of a sunspot from 2014.

Sunspots are temporary phenomena, areas on the Sun's surface that appear darker because they are cooler than the surrounding surface. They occur in areas of intense magnetic activity, which also sometimes result in large explosions of energy known as solar flares. Phenomena like sunspots are caused by changes in solar activity, which is one of the many things that scientists studying the Sun are attempting to learn more about.

Can you locate and shade in all of the sunspots in the grid showing part of the Sun's surface below, based on the clues you have been given? The sunspots are all square or rectangular areas, each at least two grid squares wide by two grid squares tall. Sunspot areas cannot touch, not even diagonally. Number clues outside the grid reveal the number of shaded squares in each row and column.

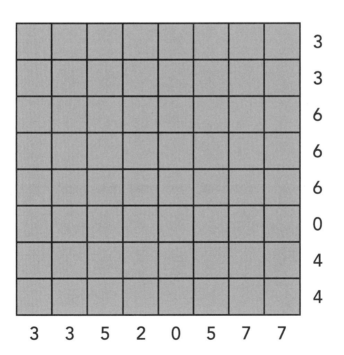

STELLAR ECLIPSE

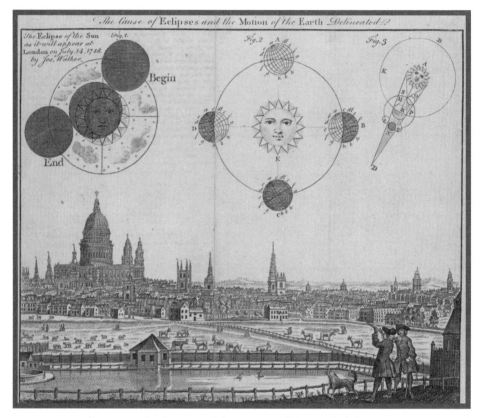

'The cause of eclipses and the motion of the earth delineated.'
This etching of London was published in the Universal Magazine in 1748,
at the time when a partial solar eclipse could be seen in the
city. The diagrams illustrate how the eclipse would appear, and
the motion of the Earth around the Sun.

The names of six bright stars have been partially eclipsed by the names of six moons, below. All of the moons are those found in our Solar System.

Can you separate the obscured celestial bodies, and work out which star and which moon is involved in each of the eclipses?

ALPIGACATETAURI

POHLEAX

ACORDELIAS

PROLTAEUS

SIYDRSA

REGOELBUS

THE PHASES OF THE MOON

The image on the opposite page shows the phases of the Moon by illustrating how it appears from Earth at each stage in its cycle, and how the phases are impacted by the orbits of the Moon and Earth around the Sun. This is one of a set of 12 astronomical prints published by James Reynolds in the second half of the nineteenth century.

The phases of the Moon below have been cut out from the original diagram on the opposite page.

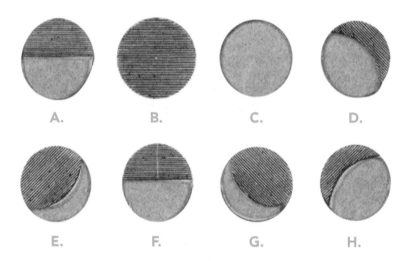

Restore each of these eight phases of the Moon to its original position in the image on the opposite page, by considering the location of the Sun and how the Moon would appear to a viewer on Earth. Each lettered Moon should be matched to a different numbered gap. The half moons are identical and so can be used in either order.

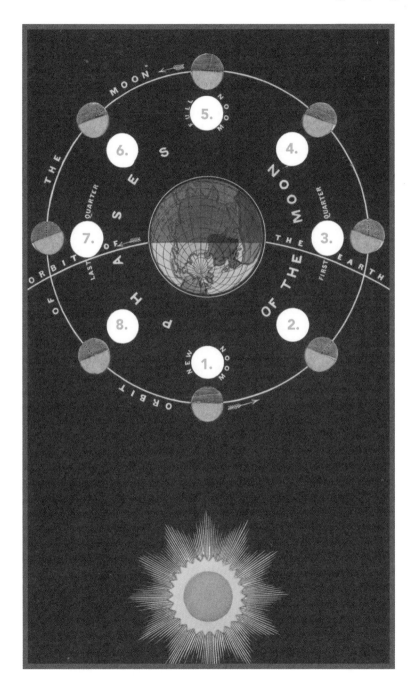

Now see if you can answer the following questions on the phases of the Moon, using the diagram on the previous page and your knowledge of the Solar System to help you:

1. Looking at the depiction of Earth on the diagram:
- What season is the northern hemisphere currently experiencing: summer or winter?
- What time of day is it in western Europe: morning or evening?

2. The two 'points' at the central edges of a crescent moon are known as its cusps, or horns. For which of the Moon's numbered phases on the diagram do the horns point eastwards? This means that they point in the same direction as the orbit of the Moon around Earth.

3. The original text accompanying the image gave the speed of the Moon's orbit around Earth as 'forty miles per minute'. Given this measurement, and supposing that it took exactly 28 days for the Moon to travel once around Earth, what would be the approximate circumference of the circle traced by the Moon's orbit around Earth.

4. The Moon takes exactly the same amount of time to rotate once around its axis as it does to complete one revolution around Earth. Also, the Earth and Moon both spin in the same direction about their respective axes, and the Moon orbits Earth in the same direction as they both spin. Given this, which of the following must be true?
- One side of the Moon never receives any sunlight.
- The phases of the Moon are caused by Earth blocking the Sun's light from reaching the Moon.
- Only one side of the Moon is ever visible from Earth.

MOON QUOTES

Astronaut and lunar module pilot Buzz Aldrin carries scientific equipment on the surface of the Moon during the Apollo 11 mission in July 1969. The equipment was left on the Moon and continued to relay information back to scientists on Earth after the departure of the astronauts. This photograph was taken by mission commander Neil Armstrong.

Each of the following quotes about Moon landings has been encoded in a different way, but all four codes share a numerical theme.

Can you crack the four codes and in so doing reveal the feature that links them all?

1. For one pricless moment n the whole hstory of man, ll the peopl on this Eart are truly on

2. Bautifu iw. Magifict dsoatio.

3. Szfdezy, Eclybftwtej Mldp spcp. Esp Plrwp sld wlyopo.

4. 30-18-11-30-29   25-24-15   29-23-11-22-22   29-30-15-26
16-25-28   23-11-24,   25-24-15   17-19-11-24-30
22-15-11-26   16-25-28   23-11-24-21-19-24-14

MOON FEATURES I

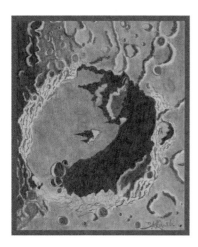

A drawing of a crater and other details on the surface of the Moon by Dr Samuel Wilfrid Russell (1895–1965). The image was created in the 1920s–1940s using a telescope in his garden in Bromyard, Herefordshire.

Features on the Moon are given Latin names, typically based on Earth-bound geographical land features that they resemble. In the grid opposite, the letters in each of the Latin names of various lunar features have been replaced with numbers, with each number consistently replacing the same letter across all of the names. Similarly, each letter is always represented by the same number. Can you crack the code and reveal the hidden Latin terminology?

The English word for each Latin feature is given, and the first line has been filled in already to get you started. (Note that some of the Latin words are also used in English)

Keep track of your deductions with this table:

1	2	3	4	5	6	7	8	9	10	11	12	13	14	15

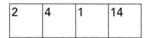

| ²M | ⁹A | ⁶R | ³E | | | Sea |

| 12 | 6 | 9 | 5 | 3 | 6 | Impact site |

| 2 | 4 | 1 | 14 | | | Mountain |

| 2 | 4 | 1 | 5 | 3 | 14 | Mountain chain |

| 11 | 4 | 6 | 14 | 13 | 2 | Ridge |

| 10 | 9 | 8 | 8 | 7 | 14 | Valley |

| 15 | 9 | 8 | 13 | 14 | | Marsh |

| 14 | 7 | 1 | 13 | 14 | | Gulf or bay |

| 8 | 9 | 12 | 13 | 14 | | Lake |

| 6 | 13 | 15 | 7 | 14 | | Rock |

MOON FEATURES II

A drawing of a crater named Arzachel on the surface
of the Moon by Dr Samuel Wilfrid Russell.

Once you have completed the previous puzzle, see if you can put your newfound lunar Latin to use. Can you unscramble the full, 'official' names of each of these lunar landscape features?

Each named feature has two parts to its name:

- Every first word can be found in the previous puzzle (e.g. MARE)

- Every second word is distinctive to an individual feature (e.g. NECTARIS – forming MARE NECTARIS).

Each scrambled name beneath contains the letters of each of its two words mixed together, but in all cases both words will contain the same number of letters.

English translations of each full name are provided too, but to make it even trickier they are not provided in the same order – so it is up to you to work out which features and which English translation go together.

Scrambled Names:

VIAL CURSES

SUMO PLAINS

MINI DISUSE

MOON PICS

TRUE SUN ATOMS

SLICK VAN PALL

English Translations:

Mountain named after the Spanish word for 'peak'

Mountain range sharing its name with a sign of the zodiac

The bay of the centre

The lake of spring

The marsh of sleep

Valley named after a Nobel Prize-winning physicist

INTERLOCKING SEAS

A pastel drawing of the Moon by the English artist John Russell (1745–1806). This highly detailed drawing was created using a micrometer to make very small measurements. Completed in about 1787, it remained the most accurate delineation of the Moon until the introduction of photography.

If you look at the Moon with the naked eye, you will notice darker patches on the surface. These are basaltic plains, formed by ancient volcanic eruptions. Early lunar observers mistook these darker areas for oceans, and the lighter areas for continents. The darker areas were named after seas (*maria* in Latin, *mare* in the singular). The names refer to sea features, sea attributes and states of mind.

The Latin names of four lunar seas have been tangled up with their four English translations, below.

Can you work out the method by which they have been combined, and then uncover all eight names for the four seas – revealing one Latin and one English name for each sea? The words meaning 'sea', in both languages, have not been included in the puzzle.

AOSTRELN

CUOIDM

FPAMING

IHBWIRM

NLBUUS

SMOREUS

SOUMANS

SUUTHARE

MOON MATCH

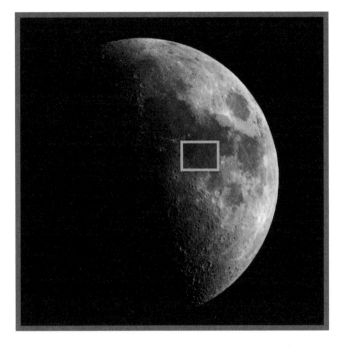

A glass lantern slide of the Moon in its first quarter, with the Sea of Tranquillity in yellow. The image was taken in May 1899.

Join up the Latin names of lunar seas on the left with their approximate English translations on the right, using straight lines to join the dots next to each name.

When joined up correctly, the lines will pass through 12 of the scattered letters which, when read in turn from top to bottom, will spell out the name of a celebrated Moon mapper.

Mare Cognitum •

H   A

Mare Frigoris •

L

Mare Humboldtianum •

U

B

Mare Humorum •

G

H

Mare Ingenii •

E   P U   F

Mare Insularum • W

M

Mare Marginis •   C   I

L Q

Mare Orientale • G

H

P   K

Mare Tranquillitatis •   I   R

S

Mare Undarum •   C

A

N

Mare Vaporum •   S

T   I

Oceanus Procellarum •

• Eastern Sea

• Ocean of Storms

• Sea of Alexander von Humboldt

• Sea of Cleverness

• Sea of Cold

• Sea of Islands

• The Known Sea

• Sea of Moisture

• Sea of the Edge

• Sea of Tranquillity

• Sea of Vapours

• Sea of Waves

WORLD FRAGMENTS

The Solar System started life as a spinning cloud of dust, before gravity pulled together rocks and gases to form celestial objects such as Earth and the Sun.

The names of seven types of celestial object – which can all be found in our Solar System – have been split up into fragments in the dust cloud below. One of the solutions consists of two words, and the others are one word long.

Can you draw the fragments together and in so doing restore the names of all of the objects? You will need to insert a space into one of the fragments.

AR	AR	AST	COM	DW
ERO	ET	ET	FP	ID
ID	LAN	ME	MO	NET
ON	ORO	PLA	ST	TE

STICKING TOGETHER

The Solar System is the gravitationally bound system of the Sun and everything that orbits it. An orbit is a regular, repeating path, which celestial objects, or satellites, take around other objects. For example, the movement of Earth around the Sun, or the Moon around Earth, are both orbits. An orbit occurs when there is a perfect balance between the forward motion of one object against the gravitational pull of another.

Can you use your decoding skills to work out which celestial objects have been drawn together in the lines below? There are two objects encoded in each line, and each pair of objects bear a certain relation to one another – but it is up to you to work out what that is, and how they have been joined together.

OU RB EA NR UO SN

PN ER OP TT UE NU SE

TS AE TT UH YR SN

IJ UO CP IA TS ET ER

CODED ORBITS

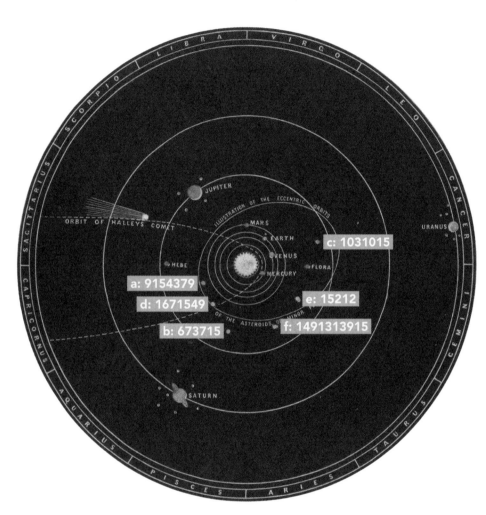

A diagram illustrating the orbits of the planets, one of a set of 12 hand-coloured astronomical prints designed to encourage learning at home. The planets are shown with their known satellites and the diagram also includes minor planets from the asteroid belt and the orbit of Halley's Comet. Published by James Reynolds in the second half of the nineteenth century.

The names of the major planets shown on the diagram opposite, as well as the name of one other celestial body, have been encoded below. Each of the letters has been replaced with a number, with the same number always replacing the same letter across the list of names. No number is used for more than one letter. Crack the number code, and then write the corresponding letter for each number in the grid below.

7 9 3 4 8

8 9 13 13 7 17 15 6 12 11 7 4

1 5 14 10 4 7 3

11 9 3 15

11 7 3 6 5 3 17

15 9 4 5 3 2

5 3 9 2 5 15

16 7 2 5 15

1	2	3	4	5	6	7	8	9	10	11	12	13	14	15	16	17

Once you have cracked the code, use the same code to reveal the names of the asteroids on the diagram, each of which has had its labels encoded with the same number code. To make it trickier, however, the spaces between numbers have been omitted – so it is up to you to decide where the spaces should be inserted in order to separate the digits into individual numbers in each coded name.

EXTRATERRESTRIAL ECLIPSES

The photos on these pages show three different eclipses in our Solar System – but unusually, none of the celestial bodies shown is our Moon. Can you work out for each image which body it is that is shown eclipsing the Sun? The names of the missions during which the images were taken are included to help you.

Image taken from the space probe Cassini *in 2006.*

Image taken from Apollo 12 in 1969.

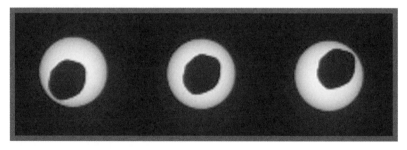

Taken from the Mars Curiosity rover in 2013. These three smaller images show the same event, three seconds apart.

MOONS OF SATURN

Can you revolve these orbital rings so that, once properly rotated, they spell out the names of five moons of Saturn? All of the letters in the inner orbit are the first letters of a moon's name, while all of the letters in the second ring out from the centre have all of the second letters, and so on.

Align the rings so that the names can be revealed, starting at the centre and reading outwards in five straight lines. All letters are used exactly once.

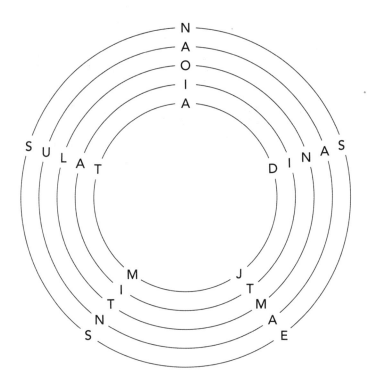

MOONS OF JUPITER

Can you revolve these orbital rings so that, once properly rotated, they spell out the names of six moons of Jupiter? All of the letters in the inner orbit are the first letters of a moon's name, while all of the letters in the second ring out from the centre have all of the second letters, and so on.

Align the rings so that the names can be revealed, starting at the centre and reading outwards in six straight lines. All letters are used exactly once.

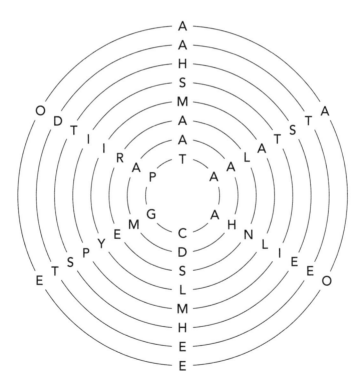

Once you have revealed the moons, can you say which one is the largest of Jupiter's moons?

CONNECTED CLUES I

Can you solve the clues below to reveal a hidden astronomical connection? All of the clues can be solved with a one-word answer, and the number of letters in each solution word has been given.

The name of the protagonist in Disney's *The Little Mermaid* (5)

Rubber disc used in ice hockey (4)

Roman god of love, often pictured as a winged boy (5)

Code word representing the letter J (6)

Pongo's mate in Disney's *One Hundred and One Dalmatians* (7)

Californian coastal city, San _____ (9)

Once you have completed the above, can you find the additional literary connection between the solutions?

CONNECTED CLUES II

Can you solve the clues below to reveal a hidden astronomical connection? All of the clues can be solved with a one-word answer, and the number of letters in each solution word has been given.

Long-standing rival (7)

The coloured part of the eye (4)

Natural phenomenon such as the Northern Lights (6)

Female friend of the protagonist in the *Harry Potter* series (8)

Lover of Pyramus who appears in the play within *A Midsummer Night's Dream* (6)

The human mind or soul (6)

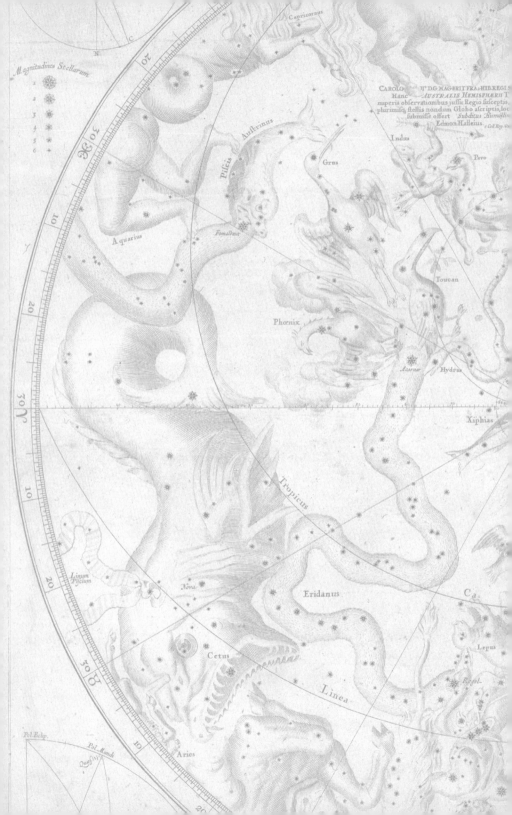

CHAPTER 4

THE SCIENCE OF SPACE

Scientists are making new discoveries about space all the time; so much so that the concepts, scope, and scale of astronomy today can sometimes feel a little overwhelming. How big is the Universe, what really happened in the Big Bang, and what exactly is the very mysterious dark energy? Put your brain to the ultimate astronomical test as you solve these puzzles about the science of space!

THE BIG BANG

Scientists believe that the Universe began approximately 13.8 billion years ago in an event known as the Big Bang. At the moment of the Big Bang, all matter and energy was encapsulated in a single point of infinite density and incredibly high temperature that suddenly exploded and expanded outwards. As the Universe continued to expand it also cooled, eventually decreasing in temperature enough for matter to form.

Can you fill in the full names of the elements created in this visual Big Bang below, by writing one letter into each gap? Only letters found in 'BIG BANG' have been given as clues.
   (In the actual Big Bang, only hydrogen and helium were formed, along with tiny amounts of lithium and beryllium.)

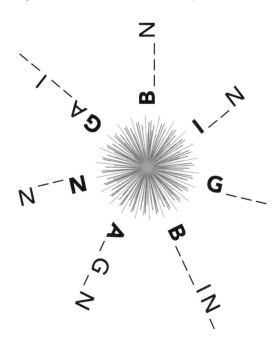

IN THE BEGINNING

Can you work out which letter comes next in each of these well-known mathematical, scientific and space-related sequences? Each of the given letters is the first letter of a word, but it is up to you to work out what those words are and therefore say what initial should come next.

1. **O T T F F S S _**

2. **S M H D W F M _**

3. **A S O N D J _**

4. **H H L B B C _**

5. **N U S J M _**

THE UNIVERSE EXPANDS

At the instant the Big Bang occurred, the Universe experienced enormous expansion in an incredibly short period of time. Afterwards, the Universe continued to grow, although the rate of expansion began to slow as it cooled. Scientists became aware of this expansion in the 1920s, when the astronomer Edwin Hubble noticed that distant galaxies were moving away from each other. Those furthest away were moving fastest, which suggested that the Universe was still expanding.

In the late 1990s, observations by the Hubble Space Telescope revealed that, instead of slowing as astronomers had predicted, the rate at which the Universe was expanding was increasing. Scientists believe this is due to a mysterious force called dark energy, but are still trying to understand its effect on the Universe.

Can you complete these astronomy-themed word pyramids, expanding each three-letter word in a series of progressions to form the final word beneath it? Each line should contain a word of the given length, using exactly the same letters in the same quantities as the word above it but plus one extra letter – although the letters will not necessarily be in the same order.

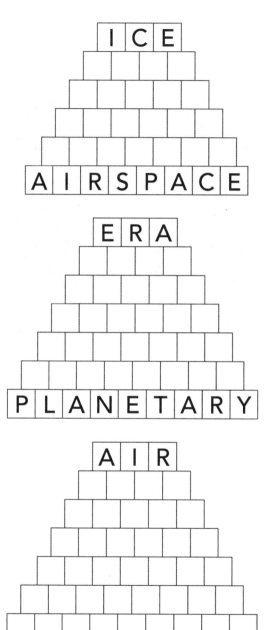

ASTRONOMICAL HOURGLASS

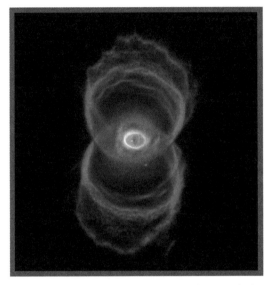

The Engraved Hourglass Nebula, approximately 8,000 light years from Earth. This image was taken by NASA's Hubble Space Telescope in 1996.

Now that you have tackled the word pyramids in the previous puzzle, can you crack the number code opposite to find what is hiding in the centre of this astronomical hourglass? Each number represents a different letter, while every line contains an English word. Each word uses the same letters as the preceding and following words plus or minus a letter as shown, although not necessarily in the same order, so can you also fill in the words in the four empty rows? Some of these empty rows have more than one possibility.

Once you have cracked the code, use it to answer the questions beneath. You will need to work out separately which letters 9 and 10 represent for questions 2 to 4.

F O R E C A S T S

6 7 3 4 1 5 1 2

— — — — — — —

2 4 3 6 5 1

— — — — —

1 2 3 4

— — — — —

2 4 5 3 8 1

— — — — — — —

3 1 1 7 4 2 5 8

A S T E R O I D S

1. Which asteroid is encoded by 6 5 4 5 1?

2. What type of vehicle is represented by the code
3 10 4 6 4 3 9 2?

3. What sub-field of fluid dynamics is encoded by
3 5 4 7 1 2 3 2 10 6 1?

4. What type of precipitation, occurring for example in the upper
atmosphere of Venus, is represented by the code 3 6 10 8 10 6?

THE STUFF OF SPACE

Although the Universe began 13.8 billion years ago, at the Big Bang, matter as we think of it was not created instantly. Matter is composed of different 'building blocks'. In the first second after the Big Bang, subatomic particles were formed, but it took more than 300,000 years of the Universe expanding and cooling before the subatomic particles began to combine to form atoms. Eventually, about 400 million years after the Big Bang, atoms combined to form the first stars and galaxies. It is inside the hearts of stars that the atoms of heavier elements like carbon and oxygen are formed.

Can you crack the code and find six subatomic particles hiding in the grid opposite? There is also an additional, more elusive entry to find somewhere in the set.

Each letter has been replaced by a number, with each number always representing the same letter. A blank grid has been provided to help you keep track of your deductions.

The first entry has been filled in already, to show how it works – and provide an initial clue.

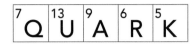

7	13	9	6	5
Q	U	A	R	K

4	9	6	5

11	9	10	10	12	6

11	13	3	1

12	15	12	14	10	6	3	1

2	8	3	1

1	12	13	10	6	3	1

2	6	3	10	3	1

1	2	3	4	5	6	7	8	9	10	11	12	13	14	15

MATTER AND ANTIMATTER

When the Universe began with the Big Bang, both matter and antimatter were created. Antimatter is the opposite of normal matter. When particles of matter and antimatter come into contact they annihilate each other. Scientists believe that matter and antimatter were created in equal quantities at the Big Bang, so the question of why the Universe today is mainly made up of matter is a mystery that physicists are still trying to solve.

The names of six subatomic particles have been replaced with their (invented) 'antimatter names' below. Can you work out how this inversion process has been applied, and then apply it again to restore the set of six original particles?

MVFGIRML

SZWILM

SRTTH YLHLM

KSLGLM

SBKVILM

HGIZMTV JFZIP

GOING ROGUE

A rogue planet is one that is not part of any particular solar system, and appears to be wandering free in outer space, away from the orbit of a star. It may have been expelled from an existing planetary system, or never been part of one in the first place.

In each of the sets of words below, there is one rogue element that is not part of the set. Can you identify the odd one out in each row?

1. Io   Mimas   Rhea   Tethys   Titan

2. Cassiopeia   Cepheus   Crux   Draco   Ursa Minor

3. Cirrus   Cumulus   Nimbus   Oort   Stratus

4. Rees   Flamsteed   Halley   Herschel   Wolfendale

5. Bellatrix   Betelgeuse   Polaris   Rigel   Saiph

'Comparative Magnitudes of The Planets' From a set of astronomical prints from the second half of the nineteenth century.

HIDDEN LIGHT

An image symbolising optics, by John Chapman, 1820.
The figures surrounding the central eye represent the seven colours
of the visible light spectrum, while the child figures in each of the corners
demonstrate optical principles and instruments: a telescope, a prism,
refraction in water and a camera obscura. The image was included in the
Encyclopedia Londinensis, *or* Universal Dictionary of Arts, Sciences
and Literature, *published between 1810 and 1829.*

Electromagnetic radiation is energy that takes many different forms. The energy travels in waves, and different types of electromagnetic radiation have different wavelengths. The range of these different types of radiation is called the electromagnetic spectrum, which is divided into seven sections. On Earth we are constantly surrounded by several types of electromagnetic radiation but our eyes detect only one – visible light. The Sun emits radiation across the entire electromagnetic spectrum, but our atmosphere forms a protective barrier against any that would be harmful to life on Earth.

Can you find the names of four electromagnetic waves hidden in the sentences below? There is one hidden in each sentence, although note that each name should be **five** letters or more.

I'd consider 'Origami Crow' a very unlikely name for a constellation

That's the postgrad I overheard talking in the corridor about the eventual collapse of the Sun

Which atoms – whose names begin 'Fr' – are decay products of uranium ores?

Don't let a minor confusion between cryptogram and cryptogam mar a year's worth of work on extraterrestrial communication

PRISM PUZZLE

The speed at which light travels changes when it moves through substances of different densities. The change in speed causes light to bend, or refract. This is why objects look different when we see them through water, for example. Refraction occurs all the time in the atmosphere, as the light from celestial objects bends as it travels through the air to reach our eyes. The different wavelengths, or colours, of visible light are refracted at different angles, causing us to see different colours in the sky. It is refraction that explains why the sky looks blue on sunny days and why we see red, orange and purple colours at sunrise and sunset.

Refraction can also be studied using a triangular glass prism. Visible light is refracted as it enters, and again it exits the prism. As the light moves out of the prism it is dispersed into the seven colours of the visible light spectrum, causing a rainbow effect.

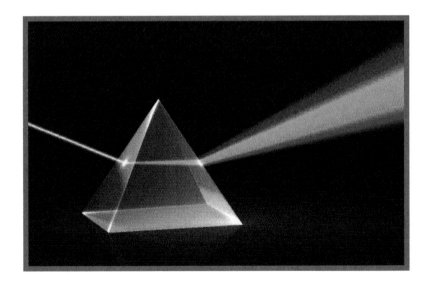

Can you place the triangular prisms into the spaces below so that each of the six rows spells the name of a colour, when read from left to right? None of the prisms should be rotated.

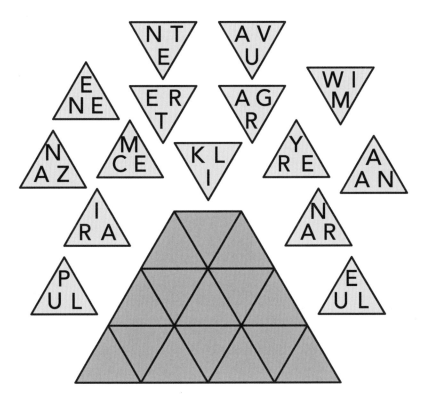

Once you have placed all the prisms, can you work out which of the revealed colours is the odd one out?

And then can you say which of these colours has a name derived from the Latin word for 'sky'?

CHEMICAL CHASMS

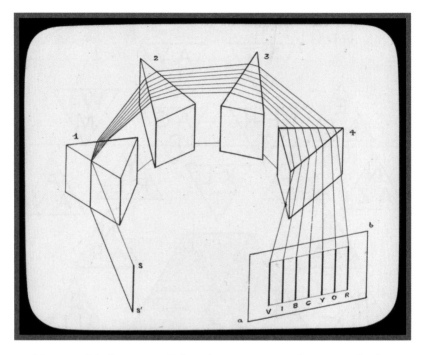

A lantern slide from a set designed to accompany a lecture on the Sun, c.1880s. This slide demonstrates the increased dispersion of light as it is refracted through a series of prisms, showing how the light can be dispersed into the different colours of the visible light spectrum.

Scientists can use a technique called spectroscopy to help determine the chemical structure of celestial objects, by looking at the way they interact with light. This works because different chemical elements absorb different amounts of light and other radiation from the electromagnetic spectrum, creating a fingerprint that can be used to help identify them. Astronomers can therefore work out which chemicals may be present in far-off places by studying the frequencies of light reaching us.

Ten chemical elements have had some important information absorbed from their names below, since several letters are missing from each element. Gaps show where letters have been absorbed from. By writing one letter into each gap, can you use your knowledge of the periodic table to fill in the gaps and identify all of the elements?

T _ _

_ _ _ C

_ _ _ _ _ T

_ Y _ _ _ _ _ _

_ _ _ _ S S _ _ _

_ _ _ _ _ _ _ Y

F _ _ _ _ _ _ E

_ _ _ _ _ _ Y

_ _ _ I _ _ N

_ _ _ K _ _

SPECTRUM SUDOKU

Solve the spectrum sudoku puzzle opposite to reveal the readings of four spectrometer plots inside the grid, and then identify the chemical elements that have been detected.

To solve the sudoku puzzle, place a digit from 1 to 9 into every empty square so that no digit repeats in any row, column or bold-lined 3×3 box. The value of the digits along each shaded colour spectrum must increase square by square from the red end (lowest value) to the purple end (highest value). This also means that digits cannot be repeated within a spectrum.

Once you have filled in all of the digits, use the data table below to identify the detected elements. Which one has been detected more than once?

SPECTROGRAM DATA TABLE	
12345	Caesium
12346	Oxygen
12347	Sodium
12356	Magnesium
12357	Argon
12456	Calcium
12457	Iron
12567	Sodium

7								4
			5					
			8		4			
		4				8	7	
	8	9				3		
			9		3			
					5			
8								3

EXOPLANETS

An exoplanet is the name given to any known planet that lies outside our Solar System. The first exoplanets were discovered in the 1990s and scientists have since found several thousand. All exoplanets are given scientific designations that are used to identify them, but some are also given proper names. The names of exoplanets and other astronomical objects are decided by the International Astronomical Union, who often invite the public to submit naming suggestions in contests called NameExoWorlds.

The five exoplanets opposite have lost their exospheres, which in this case means that the outermost (first and last) letters of their names have been removed. Can you restore these missing letters to reveal the names of five faraway worlds?

Clues to the name of each exoplanet are given, although the two lists are not in the same order so it's up to you to work out how they correspond. Note that one of the missing letters is a character found in modern Scandinavian alphabets, sometimes written as two vowels in English.

ALILE

RAH

_GI_

UIJOT

OLTERGEIS

Named after a Danish astronomer famed for his
accurate celestial observations

Named after a supernatural being whose
name translates as 'noisy ghost'

Named after an Italian astronomer who was the first to use
a telescope to see the skies

Named after the god of the sea in Old Norse mythology

Named after the noble protagonist in a Spanish
adventure novel by Cervantes

PULSAR PUZZLE

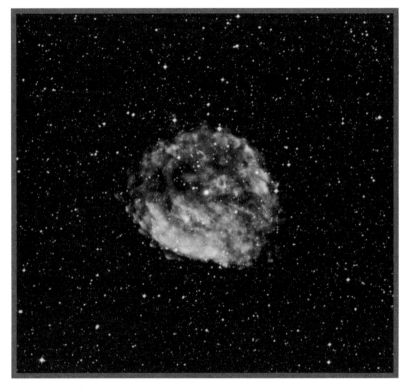

An image of 'RCW 103', captured by NASA. It is thought to be one of the most extreme pulsars, or rotating neutron stars, ever detected.

Neutron stars are formed when the outer parts of a dying supergiant star explode in a supernova, leaving behind a very small and dense collapsed core. Some neutron stars rotate rapidly and can be observed from Earth because they emit beams of electromagnetic radiation from their magnetic poles. As the star spins, the radiation is detected on Earth as a regular pulse. This type of neutron star is known as a pulsar, short for 'pulsating star', and was first discovered by Jocelyn Bell Burnell in 1967.

Not all pulsing signals come from outer space, of course, so can you use your decoding skills to discover the eight planets hidden in the pulsed signals below? Each name has been converted into Morse code, but even if you don't know any Morse code then you should still be able to solve the puzzle by using your knowledge of the Solar System to match up the planets with the various word lengths and repeated signals.

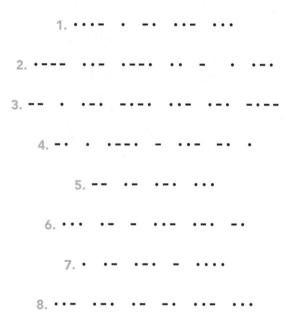

COSMIC CLOUDS

A view of the Universe produced by NASA's Hubble Space Telescope in 2014. The image was made by combining exposures taken between 2003 and 2012 and is one of the most colourful images of the Universe the telescope has produced.

Can you find your way through the cosmic clouds opposite to discover some of the secrets of the Universe? Begin at the top arrow and find a path through the three grids which travels to each letter once each, spelling out words along the way. The path can only travel left, right, up or down between squares – never diagonally – and when exiting a grid it must follow the path of the arrow shown into the next grid.

Each grid contains one astronomical term to be uncovered, consisting of more than one word. Clues in the boxes tell you which terms you are looking for in each grid, as well as the number of letters in each word.

Physicists working on this idea were looking for evidence of the origins of the universe: (3, 3, 4, 6)

T	H	N	G
B	E	A	T
I	G	B	H

O	R	E	L	E	Y	R	O	E
M	T	C	I	A				
A	G	A	D	T				
E	N	R	O	I				
T	I	C	N		C	O	E	B	A

In 1965, scientists Penzias and Wilson detected excessive noise on their instruments, which could be measured as coming from every direction in the universe. They had discovered evidence of waves which were types of: (15, 9)

The waves they found can be explained as leftover energy created when the universe was born, and their presence is known as the: (6, 9, 10)

C	O	E	B	A
M	S	V	K	C
I	W	A	G	R
C	O	R	U	O
M	I	C	N	D

BLACK HOLE DETECTOR

When a star collapses in on itself it may become a black hole, a mass so dense that it pulls everything into it, including light. Given that light cannot escape from them, black holes cannot be directly seen, except as a silhouette against the rest of space. Instead, the presence of black holes can only be inferred from measurements taken in the areas around them. In fact, there's a supermassive black hole at the centre of our own galaxy, the Milky Way – but luckily it's 26,000 light years away from the Solar System.

See if you can you use your logic skills to detect the hidden black holes in the map of part of the Universe below, based on the clues you have been given. Specifically, mark black holes in some of the empty squares, with no more than one black hole per square. Number clues reveal the total number of touching squares that include a black hole, including diagonally touching squares.

		2			2	2
3	4	3		2		
			1			
3	5		2			2
						3
3	5		4	4		
		2			3	2

DOUBLE TROUBLE

Occasionally two black holes can be found orbiting one another, in what is known as a binary black hole. Although black holes cannot be directly seen due to the fact that they swallow light, a collision of two black holes can produce an enormous explosion of heat and light, which *is* directly visible.

See if you can you use your logic skills to detect the hidden black holes in the map of part of the Universe below, based on the clues you have been given. Specifically, mark black holes in some of the empty squares, with no more than **two** black holes per square. Number clues reveal the total number of black holes in touching squares, including diagonally touching squares.

	4		3		3	
		3		2		
2		1			8	
	2		3			4
1		2		7	7	4
	5		4			2
		4		4		2

DOUBLE TROUBLE AGAIN

All of the following words have encountered a 'double black hole'. That is, the double letters that once appeared in their names have all been swallowed. Clues to their meanings are given on the opposite page, although the two lists are not in the same order – and note there are intentionally two phases of the Moon.

Can you work out which double letters have been swallowed from each word? The missing letters, when read once per pair in order, will spell out the surname of a contemporary astrophysicist, whose name in turn contains two sets of double letters.

Hule

apog

eipse

sateite

gious

vacm

oery

aular

perig

fu

Haey

At the centre of the spiral galaxy Messier 81 is a black hole that is 70 million times larger than the Sun.

- A planet-orbiting body

- American astronomer who gives their surname to an orbiting telescope

- Mechanical model of the Solar System

- Name of a comet, or its eponymous astronomer

- Oval shape, like an orbit

- Phase of the Moon

- Phase of the Moon

- Point where the Moon's orbit is closest to Earth

- Point where the Moon's orbit is furthest from Earth

- Space without matter

- Type of eclipse where the Sun remains visible as a ring around the Moon

SPAGHETTIFICATION

Black holes pull all matter into them because they have incredibly strong gravitational forces. The closer an object gets to a black hole, the stronger the gravitational pull becomes, up to a point at which it is impossible for anything to escape. This point from which nothing can return is known as the event horizon. Scientists believe that as an object approaches the event horizon and falls into a black hole the gravitational forces would cause an extreme stretching effect, until eventually the object was ripped apart. In this theory the object would be stretched vertically and compressed horizontally like a piece of pasta dough being made into spaghetti, so this effect is known as 'spaghettification'.

The words in the left-hand column opposite are approaching a black hole. Once they have passed within the black hole's event horizon, they become stretched and split, leaving gaps for extra letters.

Can you fill in these gaps with letters – one letter per marked gap – so that each of the new, longer words on the right-hand side provides the solution to one of the clues underneath? Note that the 'spaghettified' words and the clues are not necessarily given in the same order.

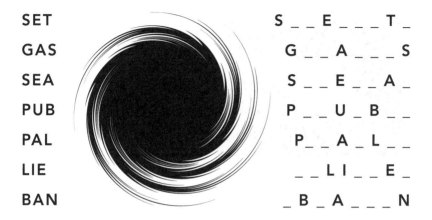

SET   S _ _ E _ _ _ T _

GAS   G _ _ A _ _ _ S

SEA   S _ _ E _ _ A _

PUB   P _ _ U _ B _ _

PAL   P _ _ A _ L _ _

LIE   _ _ L I _ _ E _

BAN   _ B _ A _ _ _ N

An orbiting body

Apparent movement when a viewpoint changes

Area of partial shadow

Erosion caused by atmospheric friction

Obscured by a celestial body

Relating to the stars

Star systems

GALAXY CROSSWORD

This image, taken by NASA's Galaxy Evolution Explorer telescope, shows one of the Milky Way's closest neighbouring galaxies.

Solve the clues to fill in the crossword, where every solution is also the name of a known galaxy – or galaxy group – in the Universe. Once complete, the letters in the highlighted squares will reveal an anagram of one our closest celestial neighbours. What is it?

Note: For 15 across, the solution is a compound word in which each half of the solution has been given a separate clue, written either side of the '+' symbol. Write them into the grid in the order given.

ACROSS

1. Tool used for hitting pucks (6, 5)

5. Spanish sun hat, perhaps (8)

6. Frog larva (7)

10. Small rodents (4)

12. Aerials; sensory appendages (8)

14. Maelstrom (9)

15. Snake-haired Gorgon + company amalgamation (6, 6)

16. Colourful, large-winged insect (9)

DOWN

2. Celestial object, usually with a tail (5)

3. Tobacco cylinder for smoking (5)

4. Tool for embroidery (6)

7. Pyrotechnic displays (9)

8. US term for a child's beach windmill; also a spinning firework (8)

9. Very tall plant with large, yellow petals (9)

11. Acrobatic sideways revolution supported with the hands (9)

13. Facial injury, perhaps as the result of a punch beneath either brow (5, 3)

SPIRAL GALAXIES

An artist's impression of the Milky Way shows its two major spiral arms.

Most known galaxies exhibit symmetry in their structure, often in more than one dimension. Our own galaxy – the Milky Way – consists of two major spiral arms, forming a large, symmetrical disk that rotates around a black hole.

Can you map out the spiral galaxies on the map of part of the universe below, so that each is symmetrically shaped around its own black hole?

Draw along some of the dashed grid lines to divide the grid into a set of regions. Every region must contain exactly one black hole, and the region must be symmetrical in such a way that if rotated 180 degrees around the black hole, it would look exactly the same. Every square is contained in exactly one region.

One of the regions has been drawn in already, to show how it works.

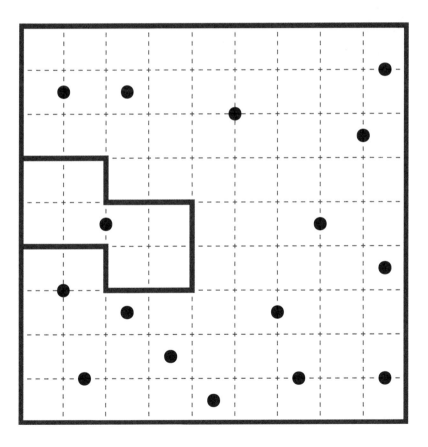

STAR SEARCH

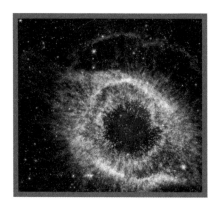

A nebula located in the Aquarius constellation, one of the closet planetary nebula to Earth.

Nebulae are enormous clouds made of dust and gas that lie in the spaces between stars. Some nebulae are the birthplaces of new stars, which is why they are often called star nurseries, while others are formed from the remnants of supernovae, or exploded dying stars. It is sometimes possible to spot the closest nebula from Earth with the naked eye, and many more can be seen through a telescope. To us, they appear as faint clouds around the stars in the night sky.

Hidden in the letter cloud opposite are the names of 12 nebulae, which may be written in any direction, including diagonally. Clues to these names are provided below. Can you find all 12 nebulae?

1. Circle; make a call (4)

2. Constellation with a famous belt (5)

3. Corkscrew shape (5)

4. Moscow square by the Kremlin (3, 6)

5. Large seabird with an expandable bill (7)

6. Nocturnal bird (3)

7. Pearl-creating mollusc (6)

8. Pelt of a vulpine animal (3, 3)

9. Decorative ribbon awarded to winners (7)

10. Saltwater lake near the sea (6)

11. Small, social insect (3)

12. Weapon which returns to the thrower (9)

```
V L Q I C J V X W V G E R
O I U Z V F Q N K N R O F
L A G O O N E Y A A Z Y F
F L Q W F T K R U Q I U O
E B V J T W E Q L L N M X
X U A E X M S X O N D W F
R S S O O D N N U P W P U
U O C O E A O R A T K C R
R S B R C I S E L N A Y N
I J D I R N M I A Q T W F
F N L O A K F P A W A O G
H E L I X O Y S T E R W D
P N M J A J V R I N G L D
```

STARRY START

Across these pages are seven images of nebulae, which have all been named after an animal that they are said to resemble. By using either your astronomical knowledge or your visual sense, can you match each of the names below with the nebula image to which you think it applies?

BUTTERFLY

CAT'S EYE

CRAB

EAGLE

HORSEHEAD

STINGRAY

TARANTULA

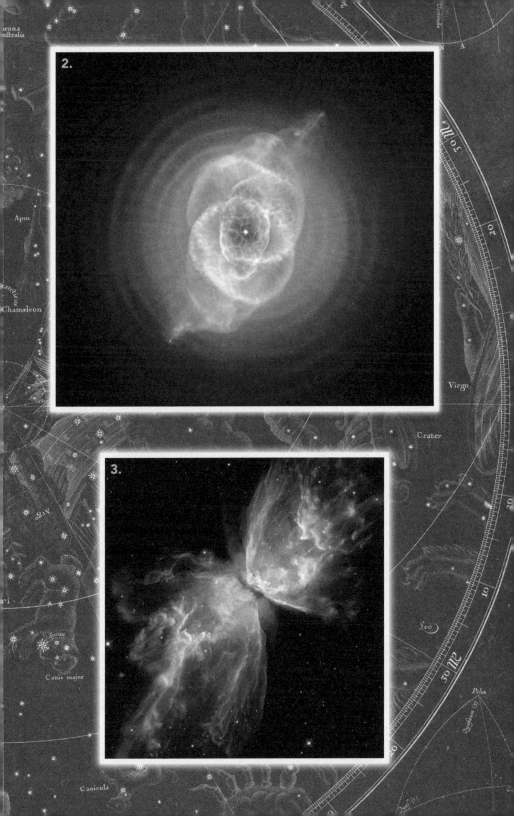

4.

5.

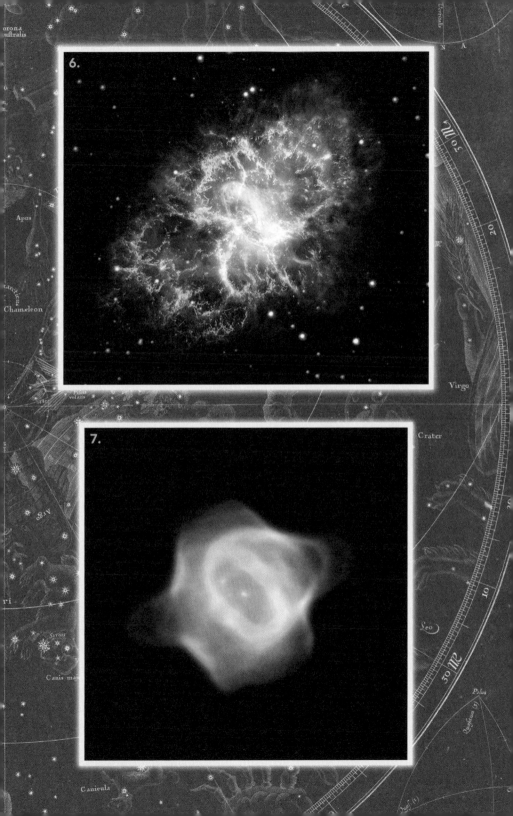

6.

7.

Magnitudines Stellarum

Capricornus

CAROLO JI D.G.MAG.BRIT FRA+HIB.REGI SE
Hanc AVSTRALIS HEMISPHÆRII Ta
nuperis obſervationibus juſſu Regio ſuſcepta, x
plurimiſq ſtellis nondum Globo aſcriptis, locu
ſubmiſſe offert Subditus Humiſſimi
Edmon Halleius. e Col Reg Oxon

Indus

Pavo

Piſcis
Auſtrinus

Grus

Toucan

Aquarius

Fomalhaut

Phœnix

Achernar

Hydrus

Xiphias

Tropicus

Eridanus

Linum
Piſcium

Ca.

Nova

Lepus

Cetus

Ridol.

Pol.Eclip.

Pol.Mundi

Aries

Linea

CHAPTER 5

NOW AND BEYOND

Today humans know more about outer space than ever before. We have watched people walk on the Moon, sent a space probe to orbit a comet and even landed rovers on Mars, but there is still far more that we have yet to understand. The puzzles in this chapter celebrate the incredible achievements made in space exploration in recent years, ponder some of the latest astronomical theories, and consider where space science may be heading in the future.

170

SPACE ON SCREEN

The names of the space-related films below have had all of the letters of the word 'ALIEN' removed from their titles, along with any spaces and digits. Can you restore the correct titles, using their brief plot descriptions as clues? The dates of their release have also been given to help you.

GRDDYOUT (1989)
A man and his dog travel to the Moon to find out whether it is made of cheese

FRSTM (2018)
A biopic following the life of Neil Armstrong and his walking on the Moon

SPCODYSSY (1968)
In this film, an alien monolith is affecting human evolution

GRVTY (2013)
Two astronauts left stranded by a collision must find their way back to Earth

TRSTR (2014)
A wormhole is explored as a last resort for the faltering human species

HDDFGURS (2016)
This biopic explores the story of three female, African-American mathematicians working in 1960s NASA

RRV (2016)
A linguist is enlisted to communicate with extra-terrestrial visitors

THMRT (2015)
A man must survive after being left alone on the red planet

POO (1995)
Three astronauts must return their spacecraft to Earth after a mechanical failure

INITIALLY HEARD

Can you restore the names of these eight pieces of music with 'astronomical' titles, where only the initials of the titles and artists have been given? Six are popular songs, while one is an orchestral movement and another is an album – but it is up to you to work out which is which.

A S C T by **C D B**

R M (I T I G T B A L L T) by **E J**

S by **D B**

T D S O T M by **P F**

J by **G H**

F M T T M by **F S**

T E O T H by **B T**

H C T S by **T B**

Once you have restored the titles, can you say which is part of a set of seven related pieces?

For a musical bonus, which astrophysicist is also the guitarist in a famous rock band, whose former lead singer had a surname with an astronomical connection?

WOMEN AND SPACE

Can you find the missing surnames of eight astronomical pioneers in the grid opposite?

To find the surnames, draw a continuous path that travels horizontally or vertically from letter to letter to spell out their surnames in the order they are clued below. The number of letters in each surname is indicated by the underlines.

No square can be entered more than once, and the path should start at the yellow star and end at the arrow leading to the red star.

1. Valentina _ _ _ _ _ _ _ _ _ _ – first woman in space

2. Svetlana _ _ _ _ _ _ _ _ _ _ – first woman to perform a spacewalk

3. Sally _ _ _ _ – first US woman in space

4. Helen _ _ _ _ _ _ _ – first British woman in space

5. Caroline _ _ _ _ _ _ _ _ – thought to be the first salaried female scientist, her brother was also a famous astronomer

6. Katherine _ _ _ _ _ _ _ – NASA mathematician who made key orbital calculations for the Apollo 11 space flight, and shares her surname with a NASA Space Center

7. Maria _ _ _ _ _ _ _ _ – US astronomer with a comet, which she discovered in 1847, named in her honour

8. Jocelyn Bell _ _ _ _ _ _ _ – astrophysicist who discovered radio pulsars

E	R	H	K	O	V	A	S
T	E	S	I	M	N	O	A
★	L	L	T	E	L	S	V
★	↵	E	C	H	L	N	I
H	E	N	R	U	B	H	T
N	R	S	C	H	E	O	S
A	A	H	D	I	L	J	K
M	R	S	E	R	A	Y	A

DOPPLER EFFECT

This space-based observatory was launched in 2002 and orbits Earth, providing scientists with a unique viewpoint from which to study space. It is named after one of the famous twentieth-century astronomers hidden in this puzzle.

When astronomers study the light emitted from distant objects in space, what they see is impacted by the Doppler Effect. This causes light to shift towards longer or shorter wavelengths as objects move relative to each other. Because the Universe is expanding, the space between Earth and distant galaxies is constantly growing, so the light emitted from those galaxies appears to have longer wavelengths. It is shifted towards the red end of the spectrum, an effect known as Red Shift. This concept has been vital in helping scientists explain the Big Bang and the expansions of the Universe.

Each of the letters in the names below has been transformed by the Doppler effect – its frequency has increased, and it has been joined by a red-herring letter. Can you delete the incorrect letter in each of the following pairs to reveal the surnames of eight twentieth- and twenty-first-century astronomers and astrophysicists?

As an example, SB CT DA RE would solve to STAR (by deleting the B, C, D and E respectively).

CE IU RN ES TI ES TI NO

HB EA WE TK EI NH GA

DC EO XB

MT YO RS OE EN

SH UY TB BA RL QE

PS TA GN VA NE

IH NA RC OT MR RA EN NC

MW RO OT NR EG

HIGH FLIGHT

The launch of the space shuttle Columbia *on 12 April 1981. This was the first orbital spaceflight of NASA's space shuttle programme.*

The poem *High Flight*, by John Gillespie Magee, is indelibly associated with astronauts and aviators. Astronaut Michael Collins took a print of it into space with him on the *Gemini 10* mission, and former NASA flight director Gene Kranz quoted it in his autobiography. Its famous closing line is often used in eulogies for astronauts, while the entirety of it is engraved on the back of the memorial to the *Challenger* disaster at Arlington National Cemetery, just outside Washington DC.

In the reproduction of the poem below, one letter has been removed from each line. Collect these letters, then rearrange them to form the name of a famous astronomer whose work provided some of the foundations for Newton's theory of gravity. The letters removed from the first verse anagram to his first name, and those in the second verse to his surname.

Oh! I have slipped the surly bods of Earth
And danced the skis on laughter-silvered wings;
Sunward I've climbed, and oined the tumbling mirth
of sun-split cluds, – and done a hundred things
You have not dreamed of – wheeled and soared and swug
High in the unlit silence. Hov'ring there,
I've chased the shouting wind long, and flung
My eager craft through footless alls of air . . .

Up, up the ong, delirious, burning blue
I've topped the wind-swept heights with asy grace.
Where never lar, or even eagle flew –
And, while with silent, lifting mind I've tod
The high untrespassed sanctity of sace,
– Put out my hand, and touchd the face of God.

SPACE AGENCIES

There are currently 72 government space agencies in existence across the globe, 14 of them with launch capability. Shown below are the English-language acronyms used by five of these national or supranational space agencies. All of these agencies maintain full launch capability.

Can you identify each agency's country or geographical region, and say what each letter stands for in each abbreviation? Note that one of the letters is taken from the second letter of the word it is shortening.

CNSA

ESA

ISRO

JAXA

NASA

Once you have identified the agencies above, can you also name two more space agencies that maintain full launch capability?

ONE SMALL STEP

Can you make the small steps – or giant leaps – between these word pairs to reveal hidden astronomical terms?

Each of the pairs of words below conceals a third word, which can be added to both the end of the first word and the beginning of the second word to create two new English words. For example, OVER _ _ _ _ KEEPER would conceal TIME, since it would make OVERTIME and TIMEKEEPER.

Can you find all four hidden words in the pairs below, each of which has an astronomical theme? The underlines indicate the number of letters in each hidden astronomically related word.

1. IMP   _ _ _   FIELD
2. FIRE   _ _ _   AWAY
3. WONDER   _ _ _ _ _   CAPES
4. FORT   _ _ _ _ _   FALL

In this second set of puzzles, the two hidden words revealed by each pair of puzzles must be combined in the order given to make one longer, single-word astronomical term. For example, if the two words were SIDE and REAL they would combine to make SIDEREAL.

5. HONEY   _ _ _ _   BEAM
     LIME   _ _ _ _ _   WEIGHT
6. AIR   _ _ _ _ _   CRAFT
     SLEEP   _ _ _ _   OUT
7. SUPER   _ _ _ _   BOARD
     SAW   _ _ _ _   PAN
8. AFTER   _ _ _   BATHE
     HOT   _ _ _ _   LESS

SCHEDULED LANDING

The image on the opposite page shows a map of the Moon, with five possible landing sites marked with letters. The rocket symbol is the current position of a space shuttle belonging to a rival space agency that is trying to land their shuttle on the moon before yours.

 The rival craft's flight path has been secretly encoded in the note below, so can you use your best detective skills to work out which of the five landing sites they are going to use to land on the Moon? You will need to make use of the grid square overlaid on the Moon.

Project Euphoria is go. Superb work so far and we have a bright future ahead of us. This is going to be downright legendary. A couple of days and we'll be landing on the Moon, barring any hold-ups. The team has a month's worth of supplies so there'll be plenty left in case of emergency. Planning to land near the major geological cleft, but don't tell anyone else.

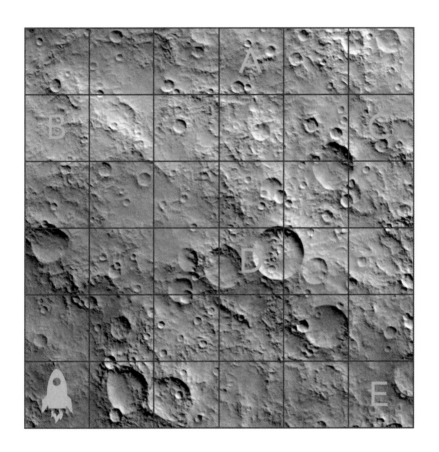

SPACE FOOD

On the page opposite are images of some of the types of dehydrated food package that were eaten by astronauts on NASA *Gemini* missions in the 1960s. Food packaged in this way needed to be rehydrated with a water gun before it could be eaten, which can be seen in the image below.

   The four foods opposite are all labelled below, but in the chaos of take-off the letters in their names have got jumbled up. Can you work out from the picture and the anagrammed labels below which foods are shown? The number of letters in each word is given. Ignore the spacing and punctuation in the jumbled versions, which is there just for readability.

1. FRY: EVADE BANG (4, 3, 5)

2. SWEETLY RARER CRAB (10, 6)

3. SPACE, EH? (7)

4. FINE, BAD CHEWS (4, 8)

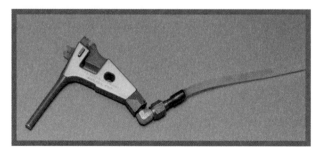

Water gun used for rehydrating food
for astronauts in space.

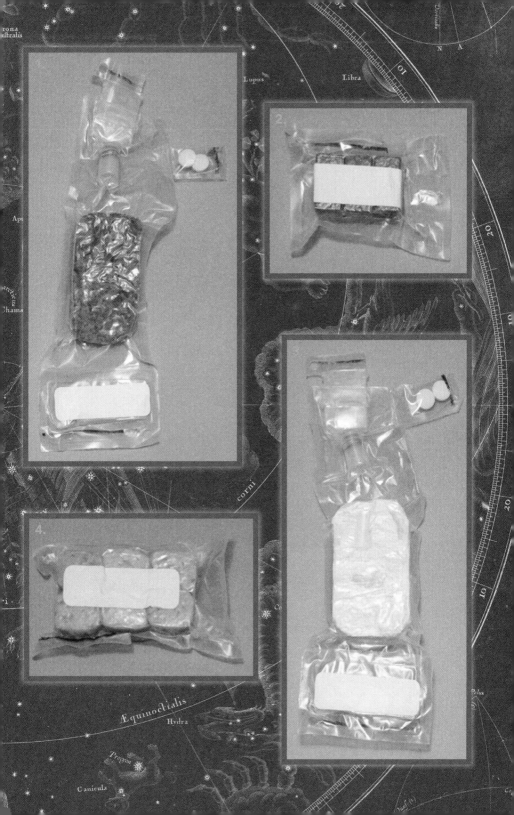

SPACECRAFT

The maiden launch of a new orbiter in NASA's Space Shuttle programme in May 1992.

Can you transform the following terrestrial vehicles into spacecraft by mixing in one of the given sets of letters?

Every vehicle in the first list can be mixed into one of the sets of letters in the second list to reveal the name of a specific spaceship (or two spaceships, in one case), such as the particular Space Shuttle shown above.

The lettersets are given in the correct letter-by-letter order for the target spaceships, but the letters in the vehicle names may need to be rearranged when added. The two lists are also not given in the same order, so it is up to you to work out how they pair together.

Vehicles

- CAB

- CAR

- TUKTUK

- VAN

Lettersets

- EDEOUR

- HLLENGE

- OLUMI

- SPNI 1 and SPNI 2

Once you have revealed the names, can you match each spacecraft to one of the descriptions below?

1. First ever artificial satellites

2. NASA shuttle whose name was chosen by children

3. Shared its name with a famous circumnavigatory ship

4. Shared its name with the lunar module that flew with *Apollo 17*

THE QUADRANGLES OF MARS

The Sojourner *rover on the surface of Mars during NASA's first Mars rover mission, Pathfinder, in 1997.*

For the purposes of efficiently mapping the planet, the surface of Mars has been divided by the United States Geological Survey into thirty quadrangles, with each quadrangle given a distinct name based on existing names chosen by early astronomers. Some of the names therefore refer to the topographical features that can be found in that particular area, while others are more whimsical.

The names of nine of these Mars quadrangles have been overlaid on the picture of Mars below, which has then been cut into nine pieces and those pieces shuffled around. Can you mentally rearrange the pieces to reveal the names of these nine quadrangles? Note that the names have not necessarily been placed anywhere near their actual location on Mars.

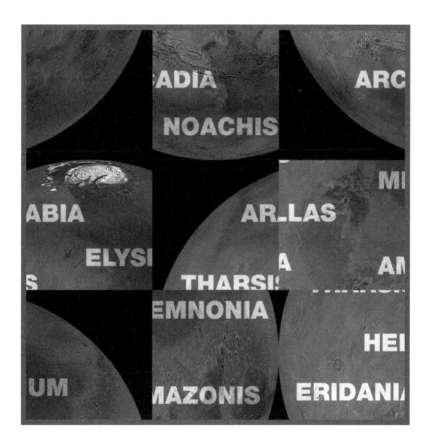

LOST IN TRANSLATION

The *Rosetta* space probe, launched in 2004, was the first probe to orbit a comet successfully. On board is a nickel engraving containing a piece of text translated into 1,500 languages, as a prototype part of the (similarly named) Rosetta Project. In theory, like its namesake stone, the engraving could be used to translate between those 1,500 languages.

Below are the names of five other space probes, each of which has travelled to distant parts of the Solar System. Meanwhile, the symbols opposite belong to an imaginary language that uses pictograms to spell English words, with each line providing a translation of one of the named probes below – although the probes are not necessarily given in the same order.

Can you work out how the pictogram system works? Note that not every individual symbol maps onto exactly one letter, so it is up to you to work out how the system fits together. Once you have cracked the code, you will be able to identify which pictogram line matches with each probe name.

JUNO

SPIRIT

ROSETTA

CASSINI

VOYAGER

CURIOSITY

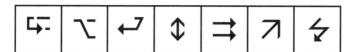

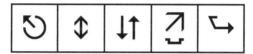

TIME TO REFLECT

This image of a famous space telescope orbiting Earth was taken during a servicing mission in May 2009.

Telescopes work by gathering light from the sky and directing it to our eyes. Reflecting telescopes do this using mirrors. The Newtonian reflector, a simple telescope design, bounces light off a curved mirror first, and then an angled flat mirror is used to direct the light towards the eye piece.

Space telescopes also use mirrors, although they need to be much bigger to observe distant objects in space and can often measure several metres across. Space telescopes are not affected by the Earth's atmosphere like those on the ground, so they can gather very clear images and are extremely useful for scientists studying the Universe.

Can you place mirrors to reveal the names of four major space telescopes, in the puzzle below? Each telescope may be either active or retired.

Draw diagonal lines between opposite corners of some squares to form mirrors, with exactly one mirror placed per bold-lined region. The mirrors must be arranged in such a way that a laser fired horizontally or vertically (as appropriate) into the grid from each lettered clue would exit the grid at another location labelled with the same letter, having bounced off the exact number of mirrors shown by the number next to the letter.

Once complete, take a look at the path of each laser beam. Read in one of the two possible directions, each beam passes through the letters spelling out the name of a space telescope. Can you uncover all four? As a clue, mirrors need not be placed in the same squares as letters.

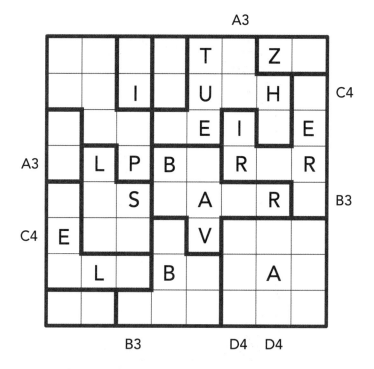

FLY-BYS

As space technology advances, we are slowly starting to see extraterrestrial surfaces in ever more detail. Breakthroughs in engineering mean that space telescopes continue to improve in clarity, and distant probes are able to send back sharper images of Solar System objects.

On the following pages are four images taken by spacecraft during fly-bys of planets. Can you work out which planet has been photographed in each image?

After you have identified the planets, can you work out which one of these images shows a planet's northern pole?

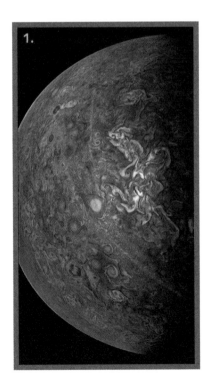

ALIEN ANNOUNCEMENT

You are a linguist who must decode the signals from the alien planet. After several weeks of work, you have managed to transcribe the following enigmatic text:

Salutatory always, fearsome

conqueror – plaintext

It appears to combine a greeting with perhaps a hint at a warning? And what is concealed in plain text?

Shortly afterwards, you receive another transmission, this time consisting entirely of 'X's and 'O's:

OXXXOOXXXO OXOOOX

OXXOXOOX XOXXOXOXX

OOOXOXOXO

What could it mean?

Take care – the future of humanity could be on the line!

MARTIAN MANAGEMENT

You are an aerospace engineer working on the development of a human colony on our nearest planet. Can you design a network of airtight bridges to link each of these colony domes on Mars? Each of the domes has been assigned a number related to its population, and so the total number of its connected bridges must be equal to the given number.

The domes are marked on the blueprint below as circles, which you should join with horizontal or vertical lines to form the bridges. No more than two bridges may join any given pair of domes, and for safety no bridges can cross over one another or over any dome.

The finished layout must of course connect all domes, so a colonist could travel between any pair of domes by using one or more bridges.

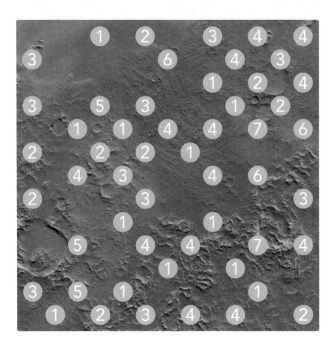

WORMHOLE WAYS

An artist's conception of a spacecraft travelling through a wormhole.

One of the greatest challenges to space exploration is the vastness of space. It took 9.5 years for the *New Horizons* mission to reach Pluto, one of the furthest points of the Solar System. But the theoretical existence of wormholes may help to solve this problem. Wormholes act a bit like shortcuts through space, tunnels that connect points which would otherwise be far apart. It may even be possible to use them to travel through time as well as space.

Can you use the wormholes below to your advantage as you travel through time and space to navigate your way through this maze?

Make your way from the entrance at the top to the exit at the bottom, warping as required between wormholes marked with identical letters that connect you to another location in the same horizontal plane. You need not use every wormhole, and you can pass over them instead of travelling through them if you wish.

THE BIG RIP 1

The Big Rip theory centres around the eventual fate of the Universe, theorising that there will come a time when the expansion of the Universe will reach infinity. The result of this will mean the Universe is ripped into oblivion, when even time will cease to exist.

Can you 'rip up' the map of the Universe below, according to the masses of celestial objects found within it? Numbers in the grid indicate the mass of certain parts, which in turn determine the boundaries along which rips can occur.
    To mark where the rips should be, draw along some of the dashed lines to divide the grid into a set of rectangular and square regions, so that each region contains exactly one number. The number in each region must be precisely equal to the number of grid squares it contains.

THE BIG RIP 2

The names of eight types of galaxy have been ripped in two. Can you join each half from the first section with a half from the second section to restore the eight galaxies to their former states?

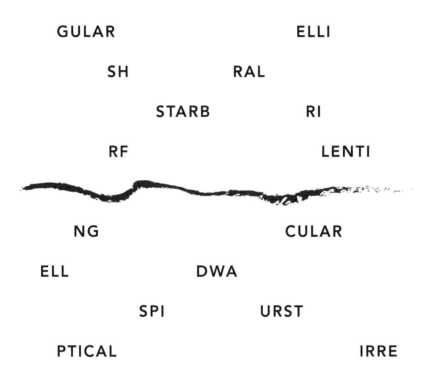

GULAR   ELLI

SH   RAL

STARB   RI

RF   LENTI

NG   CULAR

ELL   DWA

SPI   URST

PTICAL   IRRE

For bonus points, can you guess the nickname given to galaxy NGC 4151, based on its resemblance to a feature of the antagonist in the *Lord of the Rings* series?

HEAT DEATH

Heat death is a possible final fate of the Universe. It suggests that as time passes, the Universe will reach a state of maximum entropy, and temperature differences will decrease. Eventually, there will be no temperature variations, and so nothing to exploit in order to do any work (even on an atomic level). There will be no more 'progress' and time will become meaningless.

The words in the list below have all undergone 'heat death' – that is, the 'heat' has been removed from them. Can you restore the words by adding in one each of the letters 'H', 'E', 'A' and 'T' – in that order – to each set of letters so that the resulting word then matches one of the clue descriptions given on the right? For example, the word SITN can have H, E, A and T added in order to create 'HESITANT'.

OVRCRF

SOUT-S

LIOST

UMCTN

VIES

TRMOST

Hubble telescope sees stars and a stripe in celestial fireworks.

Able to retain moisture

Astronomical equipment designed to reflect the Sun's light in a given direction

Location of Paris relative to Greenwich

Temperature-regulating device

Uranium is the _____ naturally occurring chemical element

Vehicle powered by downward jets of air

THE BIG CRUNCH

The Big Crunch is another possible fate for the Universe. It suggests that there is sufficient matter eventually to overcome the expansion of the Universe gravitationally, causing it to collapse inwards and heat up, like a Big Bang in reverse. However, evidence suggests the expansion of the Universe is not slowing down, but actually accelerating.

Can you work out what words have been 'crunched' together, below?

Each line contains two words that have been fused into one and need to be separated again. The letters are all in the correct order for each original word, but the crunch has not led to a consistent blending of the two, so it is up to you to work out how to separate them.

AERGEA

YETERAMR

ASPEONAN

DFOERCTNAIDGHET

EPEPROIOCDH

CMEILLNENTNUIRYUM

CPHYCASELE

What connects all of the separated words?

SOLUTIONS

CHAPTER 1:

TIME AND PLACE

Navigational Tools (p.2)

The highlighted letters spell out OCCULTATION. This is an event where an astronomical body passes in front of another seemingly smaller one, thus obscuring it from view. Observations of moon occultations helped astronomers to make reliable predictions about time, location and other celestial events.

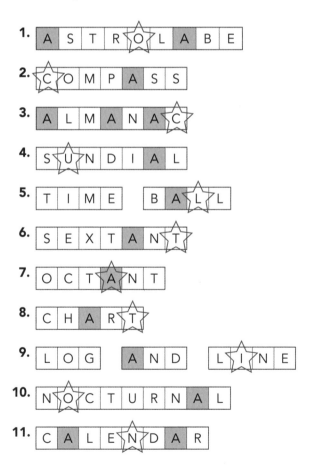

1. A S T R O L A B E
2. C O M P A S S
3. A L M A N A C
4. S U N D I A L
5. T I M E B A L L
6. S E X T A N T
7. O C T A N T
8. C H A R T
9. L O G A N D L I N E
10. N O C T U R N A L
11. C A L E N D A R

Multitasking Machines (p.6)

The observations are as follows:

- Cassiopeia was observed during harvest season, at high tide
- Scorpius was observed in the dry season, at low tide
- Leo was observed in the rainy season, at high tide

Map of Winds and Magnetism (p.8)

1. 'Ferro Island'. The label at the bottom of the map reads: 'The first Meridian from Ferro I.' Written on the map as 'Ferro I.', it can be found to the south of the Canary Islands. The island is known today as El Hierro.

2. 'The line of no variation'. The label for 'Calms and Tornados' – two opposing weather phenomena – can be found just north of the equator, in the middle of the Atlantic Ocean (labelled on the map as 'The Atlantick Ocean').

3. None at all. The island (labelled on the map as 'Bermudas') is located exactly on the line showing no variation.

4. The 20th line of longitude. Running through western France and eastern Spain, this is the line that most closely corresponds to the Greenwich meridian. The numbers indicating degrees of longitude can be found at the top and bottom of the map.

5. It is 15 degrees of east variation. The dotted line shown passing through the label for 'LA PLATA' almost exclusively runs through modern-day Argentina on the South American continent. It also passes through Brazil to the north, but appears to bypass Uruguay entirely.

6. It is 10 degrees of variation. The Azores (labelled on the map as 'Azores I.') are intersected by the line indicating 5 degrees of variation west. Tristan da Cunha, located in the area labelled 'Southern Ocean', is next to the line showing 5 degrees of variation east. The apparent difference in compass variation between the two is therefore 10 degrees.

7. Long Island, labelled on the map as 'Long I.'.

8. South-west (or south-west-by-west). The wind direction labelled 'Sept' points along the coast in a roughly south-westerly direction, pointing inland. Brazil is below the equator, marked with the thick black and white dashed line, so spring begins in September.

Prime Meridians (p.12)

The meridians, reading from left to right across the map, are as follows:

AZORES
CAPE VERDE
CANARY ISLANDS
GREENWICH
BERING STRAIT

These can be entered into the grid as follows:

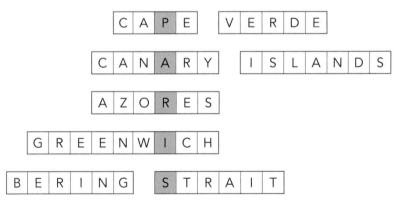

The shaded letters spell PARIS, used by France as the origin of a national prime meridian until 1911.

Prime Example (p.14)

Rearrange the meridians into this order:

9 – 5 – 12 – 1 – 3 – 6 – 2 – 7 – 11 – 8 – 10 – 4

The names can then be read as follows:

FRANCE

SPAIN

ALGERIA

MALI

BURKINA FASO

TOGO

GHANA

ANTARCTICA

Time and Tide (p.16)

The tidal descriptors, and the times they are jumbled with, are as follows:

DIURNAL + MORNING

NEAP + NOON

SPRING + MINUTE

SLACK + MONTH

EBB + DAY

FLOOD + NIGHT

HIGH + WEEK

LOW + NOW

The tide words can be classified as follows:

1. NEAP and SPRING (describing the relative size difference between high and low tides), EBB and FLOOD (describing the 'direction' of the tidal flow) and HIGH and LOW (describing tide levels).
2. DIURNAL tides are those where there is one high tide and one low tide a day.

3. SLACK tides occur when there is no tidal flow – often at the moment the tide 'turns' and changes from one direction to another.

Island Time (p.18)

1. One hour. They are 15 degrees of longitude apart, as shown by the bold lines, so they are one hour apart.

2. Point C. Being halfway between two bold lines of latitude, it is located 7.5 degrees east of the meridian, or 30 minutes ahead.

3. 60 degrees. This also makes the time difference between them four hours – one hour per 15-degree interval.

4. It is 4 p.m. Point D is 37.5 degrees to the west of point G, meaning that it is 2.5 hours behind.

5. Point G. The time is 4.5 hours ahead of the prime meridian, meaning that it must be 67.5 degrees east of the meridian. The only location at that longitude is point G.

6. D, where the time is 8 p.m. Point D is the only location 75 degrees east of point A, and the time difference is five hours.

7. F. The difference between his clock and local time is four hours, his local time being four hours ahead of the prime meridian. According to this, he must be 60 degrees east of the meridian, at point F.

8.

- 5.55 a.m. Port B is two hours behind the meridian.
- 10.25 p.m. Local time at G when he departs is 12.25 p.m. The journey takes ten hours, so when he arrives it is 10.25 p.m.
- D. Halfway through his journey is five hours in. The meridian time would be 12.55 p.m. at this point, so the clock showing 2.55 p.m. must be 30 degrees east of the meridian. D is the only location at this longitude.

Local Time (p.20)

2	2	5	2	2	1	2	2
13	5	5	5	5	7	1	7
13	13	6	6	6	7	7	7
13	13	1	6	6	6	7	3
13	13	13	13	2	2	7	3
2	2	13	13	5	5	5	3
3	13	13	3	3	3	5	5
3	3	2	2	4	4	4	4

A Subscription to Time (p.22)

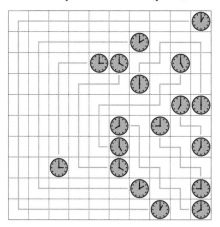

Cryptic Calendars (p.24)

Each bold number in the top-left corner of a square should be replaced with a letter corresponding to its position in the alphabet. The '1' box should therefore be read as 'A', the '2' box as 'B', and so on, until '26' is read as 'Z'.

Each set of coloured numbers encodes a different calendar name, when read in increasing numerical order. For example, for the yellow letters, the yellow number 1 is in the '1' box, so its first letter is 'A'. The yellow number 2 is in the '26' box, so its second letter is 'Z'.

The full set of calendar names can therefore be decoded as follows:
Yellow: 1-26-20-5-3 = AZTEC
Blue: 3-8-9-14-5-19-5 = CHINESE
Red: 7-18-5-7-15-18-9-1-14 = GREGORIAN
Green: 8-5-2-18-5-23 = HEBREW

Annual Anagrams (p.26)

The periods of a year are in this case all months. In the order they appear in the puzzle, they are:
JANUARY + S + W
DECEMBER + U + I
AUGUST + M + N
FEBRUARY + M + T
JULY + E + E
JUNE + R + R

The two extra sets of letters spell out SUMMER and WINTER. The changing seasons are caused by the tilt in the Earth's axis, where for parts of the year half of the planet receives more direct sunlight than the other half.

The months can then be split into two groups of consecutive months, which in the northern hemisphere correspond to summer and winter months, along with their associated season. This grouping is reversed in the southern hemisphere.

Parts of a Clock (p.27)

The unscrambled parts are as follows:
Pendulum + H
Controller + A
Escapement + R

Plate + R
Pillar + I
Balance + S
Frame + O
Pivot + N

The extra letters spell out HARRISON, the surname of John Harrison, whose clock 'H4' was the first clock that could successfully be used to keep time at sea and therefore help navigators determine their longitude. He was rewarded by the British Parliament for his role in solving 'The Longitude Problem'.

Watch the Clock (p.28)

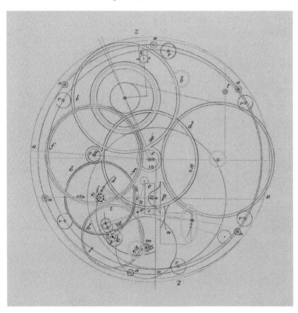

Divisions of Time (p.30)

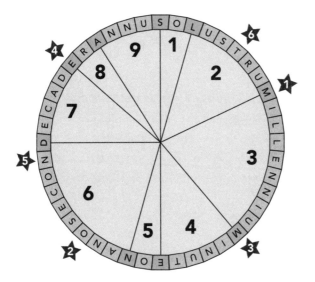

The words can be fitted into the clock as follows:
The starred letters spell out MOMENT, which was a medieval unit of time equivalent to 90 seconds.

Shakespeare's Time (p.32)

The restored lines, and the plays they appear in, are as follows:
- Better three hours too soon than a minute too late (*The Merry Wives of Windsor*, 2.2)
- We are time's subjects, and time bids be gone (*Henry IV Part 2*, 1.3)
- Let every man be master of his time (*Macbeth*, 3.1)
- I wasted time, and now doth time waste me (*Richard II*, 5.5)
- What's past is prologue (*The Tempest*, 2.1)
- There's a time for all things (*The Comedy of Errors*, 2.2)
- Time and the hour runs through the roughest day (*Macbeth*, 1.3)
- And one man in his time plays many parts (*As You Like It*, 2.7)

Time Teasers (p.34)

The images represent the following expressions:

1. All in good time
2. Behind the times
3. Time flies
4. No time like the present
5. Make up for lost time

Cloud Spotting (p.35)

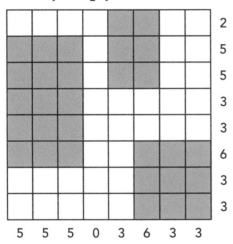

Pea-souper Letter Soup (p.36)

In alphabetical order, the five types of storm are:

BLIZZARD
HURRICANE
MONSOON
SQUALL
TYPHOON

In addition to their meteorological meanings, all of the names contain double letters.

Waves and Weather (p.37)

The labels correspond to the following phenomena:

1. The Maelstrom
2. Waterspouts
3. Fog
4. Clouds: Cirrus
5. Rain
6. Snow
7. Perpetual Snow
8. Glaciers
9. Aurora borealis
10. Rainbow
11. Halo
12. Mirage
13. Mock Suns
14. Zodiacal Light
15. Lightning
16. Falling Stars
17. Aerolites – now more commonly known as meteorites

Bonus questions:

1. Mock Suns
2. The Maelstrom
3. Zodiacal Light

CHAPTER 2:
ASTRONOMICAL HISTORY

The Figure of Astronomy (p.42)
1. STAR
2. IONS
3. NASA
4. ASTRA
5. ATOMS
6. MINOR
7. MOONS
8. ORION
9. ROMAN
10. MARTIANS

Orderly Orreries (p.44)
From the evidence given in the puzzle's timeline, and in the photo's caption, the orrery could be best estimated to have been made at any point between 1789 and 1831. There are seven planets shown, representing the first seven planets from Mercury through to Uranus. The inclusion of seven moons of Saturn suggest it was made after 1789, but the absence of Neptune suggests that it is unlikely to have been built after their discovery in 1846. In addition, the artist credited for the construction of the orrery died in 1831, so if he made it himself (rather than it being made by his studio) then it is unlikely to be later than that. In reality, it was probably made in the 1790s.

The Language of Space (p.46)
The words have the following etymologies:
COMET – From Greek, meaning 'long-haired'
CRESCENT – From Latin, meaning 'growing; to grow'
ECLIPSE – From Greek, meaning 'fail to appear'

GALAXY – From Greek, meaning 'milky'
GRAVITY – From Latin, meaning 'seriousness; weight'
MOON – From Latin/Greek for 'month', and Latin for 'to measure'
NEBULA – From Latin, meaning 'mist'
SOLSTICE – From Latin, meaning 'sun' plus 'stopped'

Extra-terrestrial Explanations (p.47)

The words match with the following definitions:
ALTAZIMUTH – B – being a combination of 'altitude' and 'azimuth'
ISOGONIC – C – meaning literally 'with equal angles'
NADIR – I – where the celestial sphere is an imaginary sphere onto which all the stars and other celestial bodies are projected
PARALLAX – F – the effect is the same reason that an object closer to you appears to move faster than an object further away
OCCULTATION – G – and thus the 'occult', meaning 'hidden'
SIDEREAL – E – also used to describe a way of measuring time, based on the apparent motion of the stars
ZENITH – H – see 'nadir' above for more details

Read Shift (p.48)

The astronomers and their achievements are as follows, in the order they appear in the puzzle:

1. **NICOLAUS COPERNICUS** – shift of 1: proposed that planets move around **one** body – the Sun
2. **JOHANNES KEPLER** – shift of 3: proposed **three** laws governing planetary orbital motion
3. **ISAAC NEWTON** – shift of 7: demonstrated that light could be split into the **seven** colours of the rainbow
4. **GIOVANNI CASSINI** – shift of 4: discovered **four** moons of Saturn
5. **WILLIAM HERSCHEL** – shift of 2: discovered **two** moons of Uranus (along with the planet itself). Herschel is the astronomer pictured.

Spider Web (p.50)

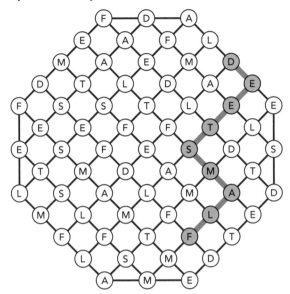

A Regular Visitor (p.52)

HAHN – surname of the 'father of nuclear chemistry' (the co-discoverer of nuclear fission, Otto Hahn)

ATACAMA – Chilean desert where the Very Large Telescope is found

CYCLICAL – occurring at regular intervals

LUCULENT – rare word meaning 'brightly shining'

SELENE – Greek goddess of the Moon

SYZYGY – the alignment of celestial bodies in straight lines, as e.g., the Sun, Earth and Moon

Starry Skies (p.54)

The rules are as follows:

Left star = constellations in the zodiac: Gemini, Virgo, Leo, Taurus

Right star = constellations named after animals: Cygnus (swan), Lupus (wolf), Leo (lion), Taurus (bull)

Overlap = zodiacal constellations named after animals

Overlapping Worlds (p.55)

The rules are as follows:

Left circle = major planets in the Solar System: Earth, Uranus, Venus, Mars

Right circle = Solar System bodies named after Roman deities: Ceres, Juno, Venus, Mars

Overlap = major planets in the Solar System named after Roman deities

Ceres is a dwarf planet and the largest object in the asteroid belt between Mars and Jupiter, while Juno is a large asteroid in the same belt.

An Epitome of Astronomy (p.56)

1. Neptune. It was not discovered until 1846.
2. William Herschel, whose name appears as 'Herschell' in the centre of the diagram underneath the name 'Saturn'. It was discovered to be a planet by Herschel in 1781. The small print originally included beneath the chart clarified that it was discovered by Flamsteed & Meyer, but they had catalogued it as a star.
3. Charts that present the Solar System as a series of concentric circles generally have the Sun at the centre, then head out towards the outer planets. On this chart, the Sun is in one of the outermost rings, with the newly discovered 'Herschel' at the centre, and the rest of the planets in 'reverse' order, conventionally speaking. It would also have made sense from the point of view of clarity to have put the smaller numbers in the smaller areas, and the larger numbers in the larger areas, so the presentation is somewhat unhelpful.
4. The distance units are given as 'English miles'. Although the length of an English mile (a 'statute mile') had been decreed by Parliament in 1593, there were still local variations, and

certainly between countries there was not yet international standardisation.

5. Venus and Mercury are shown with a difference in diameter of 5,446 miles. Venus's diameter is given as 7,906 miles and Mercury's as 2,460 miles. Mars's diameter is shown as 4,440 miles, versus Earth's of 7,964 miles, for a difference of 3,524 miles. These measurements are within 10 per cent of todays' measurements, except for Mercury, which in fact has a diameter of about 3,032 miles.

6.

- To the nearest whole number, the circumference of the Sun is approximately 10 times that of Jupiter's circumference, which is the largest circumference of the planets given. The Sun's circumference is given as 2,501,964 miles, and Jupiter's is given as 254,908 miles – making the Sun 9.815 times larger, or 10 to the nearest whole number. These numbers are within 10 per cent of modern measurements.

- Earth's circumference is almost exactly 100 times smaller than that of the Sun's, based on this chart at which it is measured at 25,020 miles compared to the Sun's 2,501,964 miles. Venus is shown with a circumference of 24,825, which is also very close to being 100 times smaller. These numbers are within 10 per cent of modern measurements.

7. The Venus solar day is shown as being one hour shorter than an Earth day. The modern estimate of the length of a Venus solar day is about 117 Earth days (and a sidereal day length of about 243 days). Venus also spins in a retrograde fashion, meaning it rotates the opposite way to Earth, which could be why the readings used for this chart were so inaccurate.

8. About 12 years. Jupiter takes 4,332.5 days to orbit the Sun, according to the chart, meaning it takes $4332.5 \div 365.25 =$

11.86 or about 12 Earth years to orbit the Sun. This agrees with modern measurements.

9. According to the distances from the Sun given, Venus and Earth can come closer than any other two planets ever do, with a difference in the radius of their orbits of 81 million minus 59 million = 22 million miles. The modern difference in their orbits is measured at 26 million miles, but this is still less than the modern difference between Mercury and Venus of 31 million miles. It is worth noting that on average, however, since Mercury orbits most tightly around the Sun, it is usually closer to Earth than Venus is!

10. 395 miles per hour. The velocity of Jupiter is given as 362 miles per minute, which is equivalent to 362 × 60 = 21,720 miles per hour. This is shown to be 55 times swifter than a cannon ball, so a cannon ball must be assumed to travel at 21,720 ÷ 55 = 394.9 miles per hour.

Constellation Computation (p.60)

Take the first letters of each group of four in turn to find the first name; then take the second letters for the second name, and so on. The constellations, and the illustrations they match to, are:

- URSA MAJOR – B
- DELPHINUS – C
- CENTAURUS – D
- ANDROMEDA – A

It is the Andromeda galaxy which is thought to be on a collision course with our own galaxy.

Illustrated Heavens 1 (p.62)

The constellations are as follows, along with their disguised form used in the aide-memoire list:

1. Pisces (PSES)
2. Aries (RES)
3. Leo (LE)
4. Taurus (TURUS)
5. Sagittarius (SGTTRUS)
6. Virgo (VRG)
7. Cancer (NER)
8. Scorpio (SRP)
9. Libra (LBR)
10. Capricorn (PRRN)
11. Gemini (GEMN)
12. Aquarius (QURUS)

Illustrated Heavens 2 (p.66)

The star images correspond to the following numbered illustrations:

A. 6: Virgo	B. 1: Pisces	C. 7: Cancer
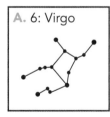	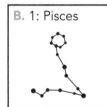	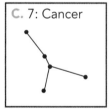
D. 8: Scorpio	E. 10: Capricorn	F. 9: Libra
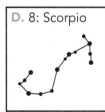	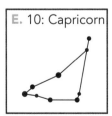	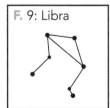

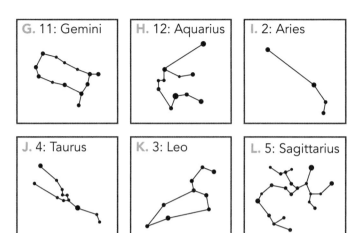

| G. 11: Gemini | H. 12: Aquarius | I. 2: Aries |
| J. 4: Taurus | K. 3: Leo | L. 5: Sagittarius |

Constellation Charts (p.68)

1. Ursa Major.

2. (The) Great Bear.

3. It is a quadrant, used to measure angles up to 90 degrees.

4. A sextant. A sextant covers one sixth of a circle (an arc of sixty degrees), whereas a quadrant covers one quarter of a circle (an arc of ninety degrees).

5. Corona Borealis. It can be found to the left of the male figure, below the compass. It shares part of its name with the aurora borealis, also known as the Northern Lights.

6. Mons Mænalus is shown at the bottom of the image and depicts a mountain on which a herdsman is stepping. 'Mons' is the Latin word for 'mountain', now used as a word in the names of mountains on planetary bodies, such as 'Olympus Mons' on Mars.

7. Arcturus, near the figure's left knee (from his point of view) is the fourth brightest star in the night sky (excluding the Sun), and is marked with a Greek 'alpha'. This symbol usually indicates the brightest star in a constellation. The additional illustrated detail around the star is an indication of its greater-than-average brightness; this same emphasis can also be seen in the illustration of the star Denebola in the bottom-right corner of the image (which is the sixtieth-brightest star in the night sky).

Southern Skies (p.70)

1. Syrius – nowadays spelled as Sirius – is marked as the brightest star on the map. It can be found in the constellation labelled 'Canis major', halfway between the bottom and centre of the map. Sirius, also known as the dog star (since it is within Canis Major, the 'great dog'), is in fact the brightest star in the night sky.

2.
- The Southern Cross. The constellation labelled *Croſiers*, now known as Crux, can be found on the chart by tracing rightwards from the curved label for 'Polus Antarcticus'.
- The countries are Australia, Brazil (among other stars), New Zealand, Papua New Guinea and Samoa.

3. A toucan, labelled in English on the map as 'Toucan'. This can be seen by tracing to the left of the southern pole star, and is also known as Tucana.

4. Corona auftralis (Corona Australis). An accompanying illustration depicts it as a crown, but it may be hard to see on the reproduction of the map. It has a 'sister' constellation – the Corona Borealis – which can be found in the northern celestial hemisphere.

5. The Equator. Orion can be found at the bottom of the chart, depicted upside down as a man holding a club. The three bright stars which make up his 'belt' are drawn close to the line labelled 'Linea Æquinoctialis' – Latin for 'equator line'.

6. A bowl, or specifically a mixing bowl. The constellation labelled 'Crater' on the chart, on the far right of the image, is depicted as a large two-handled vessel, and similarly lunar (and planetary) craters are so-named for their bowl-like structure. The English word 'crater' came from the word 'krater' in Ancient Greek, which also meant 'mixing bowl'.

7. (The) Argo. The ship is upside down in the bottom-right quarter of the chart, with its keel resting upon the centre of the map. The label 'Argo' is written on the stern of the ship.

8. Each of the small symbols represents one of the signs of the zodiac – there are twelve segments, one for each sign. The constellations pertaining to each of the zodiacal areas can be found illustrated along the outer edge of the celestial hemisphere, although they are all only partly visible.

Asterisms (p.74)

The restored names of the asterisms are:

1. Trapezium, part of the constellation of Orion.

2. Spring Triangle, an imaginary triangle visible in the northern hemisphere between March and May, joining the bright stars Arcturus, Spica and Regulus.

3. Seven Sisters, also known as the Pleiades, is the most obvious star cluster in the night sky.

4. Teapot – part of the constellation of Sagittarius.

5. Orion's Belt – part of the constellation of Orion.

6. Great Square of Pegasus – part of the constellation of Pegasus.

Celestial Deities (p.76)

The deities have all had the letters M, O and N removed – as they are all gods and goddesses associated with the Moon. Their restored names are as follows:

SELENE – Greek

DIANA – Roman

MÁNI – Norse

PHOEBE – Greek

KHONSU – Egyptian

CHANG'E – Chinese

ARTEMIS – Greek

Chang'e, a Chinese moon goddess, is also the name of China's
lunar exploration project.

Scrambled Sense (p.78)
Unscrambled words are shown in upper case:
* A small NEBULA can be seen in the 'sword' of the Orion
  constellation
* Copernicus has a lunar CRATER named after him
* Galileo was the first to DISCOVER moons other than our own
* Earth and Mars have ice caps on their POLES
* COLLAPSING stars sometimes form black holes
* A lunar eclipse won't be visible from all LONGITUDES
* The scattering of SUNLIGHT through atmospheric particles
  makes the sky appear blue
* Venus and Mars were once mistakenly considered
  WANDERING stars

Zodiac Zig Zags (p.79)
The full words are as follows:
ZODIACAL
ALKALINE
NEUTRINO
NOONTIME
MERIDIAN
ANALEMMA
MAGNETIC
ICE CAPS

CHAPTER 3:
THE SOLAR SYSTEM

Atmospheric Words (p.82)

The words which can be spelled out are as follows:

1. Atom
2. Eros
3. Eras
4. Horse (its name is Latin for 'little horse')
5. Rotor
6. Horst
7. Torero (a bullfighter – who might confront Taurus, the bull)
Bonus question: The thermosphere

Concealed Planets (p.84)

The words can be found as follows, with underlines showing the names of the hidden planets:

1. Mars: These clouds are probably going to *mar s*ome of the lunar eclipse viewings.
2. Venus: There are ele*ven us*able emergency exits on the spacecraft.
3. Uranus: The ground control burea*u ran us* through the list of spacewalk protocols before take-off.
4. Saturn: Make sure everybody get*s a turn* using the telescope.
5. Neptune: The landing made us look i*nept; une*ven lunar surfaces are not ideal touchdown sites.
6. Earth: Can you actually h*ear th*e sound of the aurora borealis?

Goldilocks and the Three Thermal Ranges (p.86)

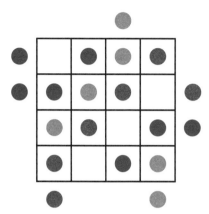

Some Sun Sums (p.88)

The numbers and measurements can be matched as follows:

A. Age of the Sun in billions of years: 4.6

B. Approximate number of minutes it takes for the Sun's light to reach us: 8

C. Diameter of the Sun, in millions of kilometres: 1.39

D. How much brighter the Sun appears to us than the next-brightest star, Sirius: 13 billion (it is worth noting that the Sun only appears to be much brighter. In terms of its luminosity, Sirius is far brighter than our Sun)

E. How many times greater the Sun's mass is than Earth's: 330,000

F. Percentage of the Sun that is composed of hydrogen: 91

G. Temperature of the Sun's core in millions of degrees Celsius: 15

Solar Eclipses (p.90)

The eclipse images pair with the diagrams as follows:

A. 1

B. 3

C. 2

D. 4

The blue lines below show the field of vision for each eclipse-watcher on Earth, for each diagram from A to D:

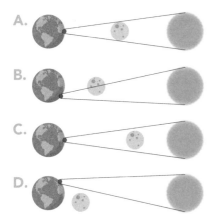

Eclipsed Dictionary (p.92)

In the order they are given in the puzzle, the completed words, and their matching definitions, are as follows:

- NUCLEAR FUSION – the process via which the Sun releases heat and light
- CORONA – outermost edge of the Sun's atmosphere, visible during a solar eclipse
- LUMINANCE – measure of an object's brightness, with SI unit candela
- MAUNDER MINIMUM – historical period of time noted for its low sunspot activity, named after a famous astronomer (Annie Maunder and her husband Edward)
- SOLAR WIND – stream of charged particles from the Sun, causing auroras
- PHOTOSYNTHESIS – the process that supports life using the Sun's rays
- SUNSPOT – dark patch on the Sun's surface
- SYNODIC DAY – the time taken for a planet to rotate around its axis once in relation to its nearest star

Eclipse Collection (p.94)

The correct terms are as follows, along with their associated usage in terms of describing eclipses:

ANNULAR – a ring-shaped eclipse

LUNAR – an eclipse of the Moon

PARTIAL – an eclipse of only a partial area of a celestial body

PENUMBRA – an area in partial shadow during a partial eclipse

SOLAR – an eclipse of the Sun

TOTAL – a complete eclipse of a celestial body

UMBRA – an area of total shadow

Bonus question: The word 'umbra' is also used to describe the dark centre of a sunspot.

Solar Deities (p.96)

The names can be entered into the grid as follows (letters in tall boxes are repeated on each row for clarity):

Grid						Culture
R	A					Egyptian
S	A	U	L	E		Baltic
S	U	É				Muisca
S	U	R	Y	A		Hindu
S	U	L	I	S		British Celtic
S	O	L				Roman
H	E	L	I	O	S	Greek
M	E	R	I			Brazilian Indigenous
I	N	T	I			Inca
E	K	H	I			Basque
X	I	H	E			Chinese
A	I	N	E			Irish

The letters in the highlighted squares are U, A, S, N and N. They can be rearranged to spell SUNNA, the name of the sun goddess in Old Norse mythology.

Sun Scape (p.98)

The grid can be filled in as follows, with the corresponding clue shown by each line:

S U N B U R S T   Rupture

S U N B O W   Front of a ship

S U N S P O T   Notice (something)

S U N L I G H T   Not heavy?

S U N R I S E   Increase, like a price

S U N B E A M   Rafter

S U N D I A L   Start a phone call

S U N D O W N   Soft feathers

A Spot of Sun (p.100)

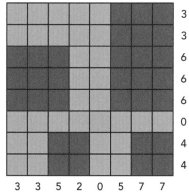

Stellar Eclipse (p.102)

The names of the moons are placed on top of the names of the stars, with common areas in black. In the order they appear, the combinations are as follows, with the stars listed first:

- Alpha Centauri and Galatea
- Pollux and Rhea
- Arcturus and Cordelia
- Polaris and Proteus
- Sirius and Hydra
- Regulus and Phoebe

The Phases of the Moon (p.104)

The moons can be reinstated as follows, matching up lettered cutouts with their numbered positions on the diagram:

A. 3 or 7
B. 1
C. 5
D. 6
E. 8
F. 3 or 7
G. 2
H. 4

1.
- Summer. The tilt of the Earth's axis is pointing the northern hemisphere towards the Sun on the diagram. When this occurs, the northern hemisphere receives more sunlight than the southern, causing the effect of summer.
- Evening. As the Earth rotates eastwards, the part of the diagram showing Europe (coloured in red) is about to rotate towards the side of the Earth that is in shadow, or night.

2. Phase 2. In this phase, where the Moon is a waxing crescent, the horns point in the same direction as the Moon's orbit, or eastwards. In the waning crescent, shown as figure 8 on the diagram, the horns point to the direction from which the Moon has just travelled, or westwards.

3. Approximately 1.6 million miles. The figure of 40 miles per minute must be multiplied by 60 (minutes) and 24 (hours) to give the distance travelled in one day, then by 28 (days). This is close to the modern measurement of almost exactly 1.5 million miles.

4. Only option 'c' is true: 'only one side of the Moon is ever visible from Earth'. For 'a', although the side facing away from Earth is known as the 'dark side' it does receive sunlight; in fact during a new moon the 'dark side' of the Moon receives full sunlight. For the situation described in 'b', that the Earth blocks light from reaching the Moon, it would need to be a lunar eclipse – which is not related to the phases of the Moon.

Moon Quotes (p.107)

1. 'For one priceless moment in the whole history of man, all the people on this Earth are truly one.' – United States president Richard Nixon, during a phone call to Aldrin and Armstrong on the Moon. Every eleventh letter has been deleted.

2. 'Beautiful view. Magnificent desolation.' – Buzz Aldrin, upon walking on the Moon's surface. The letters of the word 'eleven' have all been removed from the quotation.

3. 'Houston, Tranquility Base here. The Eagle has landed.' – Neil Armstrong, upon the occasion of the Moon landing. It has been encoded using a Caesar cipher, whereby the letters are shifted forward a constant amount in the alphabet. In this case, letters have been shifted forward 11 places, so that A = L, B = M, and so on, until Z = K.

4. 'That's one small step for man, one giant leap for mankind.' – Neil Armstrong, as he stepped onto the Moon's surface. Letters have been replaced by numbers representing their position in the alphabet, using 11 as the starting point; in this case A = 11, B = 12 and so on, until Z = 36.

All of the codes involve the number 11 in some way. This is because all of the encoded sentences were spoken during the *Apollo 11* space mission, in which mankind landed on the Moon for the first time.

Moon Features I (p.108)

The number-to-letter correspondences are:

1	2	3	4	5	6	7	8	9	10	11	12	13	14	15
N	M	E	O	T	R	I	L	A	V	D	C	U	S	P

The names in the order they appear in the puzzle are as follows:
CRATER
MONS
MONTES
DORSUM
VALLIS
PALUS
SINUS
LACUS
RUPIS

Moon Features II (p.110)

The unscrambled names are, in the order given:
- LACUS VERIS – the lake of spring
- PALUS SOMNI – the marsh of sleep
- SINUS MEDII – the bay of the centre
- MONS PICO – mountain named after the Spanish word for 'peak'
- MONTES TAURUS – mountain range sharing its name with a sign of the zodiac
- VALLIS PLANCK – valley named after a Nobel Prize-winning physicist

Interlocking Seas (p.112)

The names have been encoded by taking one letter alternately from the Latin and English name for each sea and combining them. For example, MOON and SEAS when intertwined in this way would give MEOS and SOAN. To solve the puzzle, tangled names of the same length should be paired up, and then one letter taken from each in turn until the completed names are revealed.

The names of the seas are as follows, with the Latin names on the left and English translations on the right:

(Mare) Australe	Southern (Sea)	AOSTRELN/SUUTHARE
(Mare) Imbrium	(Sea of) Showers	IHBWIRM/SMOREUS
(Mare) Nubium	(Sea of) Clouds	NLBUUS/CUOIDM
(Mare) Spumans	Foaming (Sea)	SOUMANS/FPAMING

Moon Match (p.114)

When the names and their translations are correctly matched, the following letters are intersected:

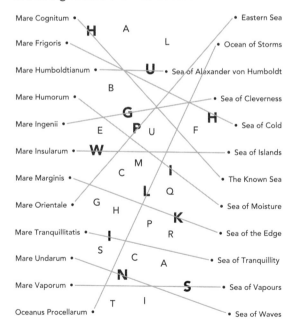

Read from top to bottom, the crossed letters spell out HUGHPWILKINS, or Hugh P. Wilkins – the creator of highly detailed maps of the Moon.

Word Fragments (p.116)
In alphabetical order, the items are as follows:
ASTEROID
COMET
DWARF PLANET
METEOROID
MOON
PLANET
STAR

Sticking Together (p.117)
Each of the lines consists of a planet from our solar system along with one of its own moons. For each line, take one letter from each pair to spell out the name of the planet, leaving the remainder of the letters to spell out the name of one of its moons.

In the order they appear, the pairs are as follows:
URANUS and OBERON
NEPTUNE and PROTEUS
SATURN and TETHYS
JUPITER and IOCASTE

Coded Orbits (p.118)
In order, the names can be decoded as follows:
EARTH
HALLEYS COMET (Halley's Comet)
JUPITER
MARS
MERCURY
SATURN

URANUS

VENUS

Which gives the following letter = number correspondence:

1	2	3	4	5	6	7	8	9	10	11	12	13	14	15	16	17
J	N	R	T	U	C	E	H	A	I	M	O	L	P	S	V	Y

This can then be used to decode the names of the asteroids as follows:

a. ASTREA (9 15 4 3 7 9) – which is now more commonly spelled ASTRAEA
b. CERES (6 7 3 7 15)
c. IRIS (10 3 10 15)
d. VESTA (16 7 15 4 9)
e. JUNO (1 5 2 12)
f. PALLAS (14 9 13 13 9 15)

Extraterrestrial Eclipses (p.120)
In the first image, the planet Saturn is eclipsing the Sun. It was seen from the *Cassini* space probe, and illuminated unprecedented detail in the composition of Saturn's rings.

In the second image, Earth is eclipsing the Sun. The crew of *Apollo 12* took this photo on their return journey home from the Moon in 1969.

In the third image, Mars's larger moon Phobos is eclipsing the Sun. The photo was taken on Mars by NASA's *Curiosity* rover.

Moons of Saturn (p.122)

The names of the moons that can be spelled out are:

ATLAS
DIONE
JANUS
MIMAS
TITAN

The rings should be arranged as follows:

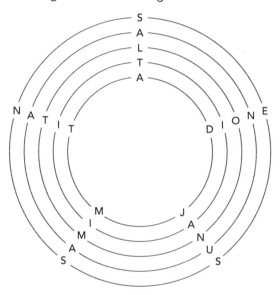

Moons of Jupiter (p.123)

The names of the moons that can be spelled out are:

ADRASTEA
AMALTHEA
CALLISTO
GANYMEDE
PASIPHAE
THEMISTO

The rings should be arranged as follows:

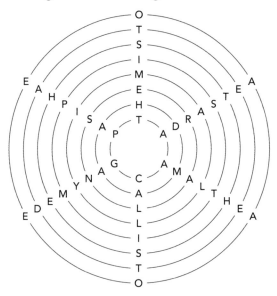

Ganymede is the largest of Jupiter's moons.

Connected Clues I (p.124)

The solutions are all the names of moons of Uranus. In the order they appear, the clues can be solved as follows:

- Ariel
- Puck
- Cupid
- Juliet
- Perdita
- Francisco

All of the names appear as characters in plays by Shakespeare – the source from which many of the moons of Uranus were named.

Connected Clues II (p.125)
The solutions are all the names of asteroids in the asteroid belt
between Mars and Jupiter. In the order they appear, the clues can
be solved as follows:
- Nemesis
- Iris
- Aurora
- Hermione
- Thisbe
- Psyche

CHAPTER 4:
THE SCIENCE OF SPACE

The Big Bang (p.128)
The names of the chemical elements, working clockwise from the
top point of the star, are:
- BORON
- IRON
- GOLD
- BROMINE
- ARGON
- NEON
- GALLIUM

In the Beginning (p.129)
The next letter in each sequence is:
1. E, for 'eight'. The sequence is positive integers in increasing
   size order: one, two, three, four, five, six, seven.
2. Y, for 'year'. The sequence is common units of time in
   increasing order of duration: second, minute, hour, day, week,
   fortnight, month.

3. F, for 'February'. The sequence is months of the year, in calendar order: August, September, October, November, December, January.
4. N, for 'nitrogen'. The sequence is chemical elements, in increasing order of atomic number: hydrogen, helium, lithium, beryllium, boron, carbon.
5. E, for 'Earth'. The sequence is planets of the Solar System, beginning furthest from the Sun and moving inwards: Neptune, Uranus, Saturn, Jupiter, Mars.

The Universe Expands (p.130)
Possible solutions (since some rows have alternative fits) to the pyramids are:
1. ICE
2. EPIC
3. PRICE
4. PRICES
5. SPACIER
6. AIRSPACE

1. ERA
2. NEAR
3. LEARN
4. RENTAL
5. PLANTER
6. PATERNAL
7. PLANETARY

1. AIR
2. LAIR
3. RAILS
4. SERIAL
5. REALIST

6. EARLIEST
7. STATELIER
8. EARTHLIEST

Astronomical Hourglass (p.132)
The two pyramids can be completed as follows, although some lines have alternate possibilities:
FORECASTS
COARSEST
COASTER
TRACES
CARTS
STAR
DARTS
TREADS
ROASTED
ASSORTED
ASTEROIDS

The object hiding in the middle of the astronomical hourglass is therefore a STAR.

This means that the number to letter correspondences are as follows:
1. S
2. T
3. A
4. R
5. E
6. C
7. O
8. D

For the additional questions, the letter 'F' has been replaced by 9 and the letter 'I' by 10. This gives the following answers:

1. CERES – the largest asteroid in our Solar System, also designated as a dwarf planet
2. AIRCRAFT
3. AEROSTATICS – the study of gases that are not in motion; it is also the name given to the study of ballooning and lighter-than-air aircraft
4. ACIDIC – although it never reaches ground level, sulphuric acid falls as rain in the outer layers of Venus's atmosphere

The Stuff of Space (p.134)

The names can be entered into the grid as follows:

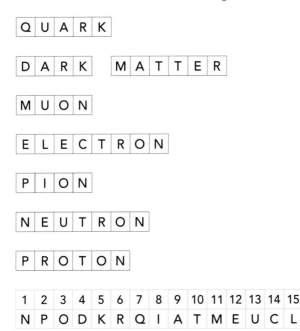

Dark matter is required to explain many astronomical observations, but has so far never been directly observed – hence the name, 'dark' matter.

Matter and Antimatter (p.136)

The names have been encoded using an Atbash cipher, in which each letter in the alphabet is replaced with the letter in the 'opposite' position in the alphabet; A = Z, B = Y and so on, until Y = B and Z = A. In this way, the letters have been replaced with their 'anti' letters.

When decrypted, the original names are:

- NEUTRINO
- HADRON
- HIGGS BOSON
- PHOTON
- HYPERON
- STRANGE QUARK

Going Rogue (p.137)

The rogue elements in each set are as follows:

1. Io – a moon of Jupiter. The rest are moons of Saturn.

2. Crux – a constellation found in the southern celestial hemisphere. The rest are found in the northern celestial hemisphere.

3. Oort – a 'space' cloud made of dust and ice, encircling the Solar System. The rest are names for water-vapour cloud formations found in Earth's atmosphere.

4. Herschel – the surname of astronomer William Herschel. His is the only name on the list who has *not* been Astronomer Royal.

5. Polaris – the North Star. The other four stars can all be found in the constellation of Orion, as the four main outer points of Orion's body.

Hidden Light (p.138)

The hidden waves are as follows:

- Microwave: I'd consider 'Origami Crow' a very unlikely name for a constellation.

- Radio: That's the postgrad I overheard talking in the corridor about the eventual collapse of the Sun.
- Infrared: Which atoms – whose names begin 'Fr' – are decay products of uranium ores? (The correct answer is Francium.) 'Infrared' also contains the word 'red', but this is not five or more letters in length as required by the question.
- Gamma ray: Don't let a minor confusion between cryptogram and cryptogam mar a year's worth of work on alien communication.

Prism Puzzle (p.140)

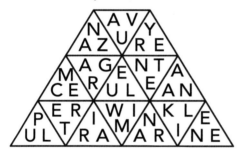

The names of the colours are:
NAVY
AZURE
MAGENTA
CERULEAN
PERIWINKLE
ULTRAMARINE

Magenta is the odd one out, since all of the other colours are shades of blue.

The word 'cerulean', referring to a deep sky-blue colour, comes from the Latin *caeruleus*, meaning 'sky blue'. The Latin for 'sky' is *caelum*.

Chemical Chasms (p.142)

The chemical elements, in the order they appear, are:

- TIN
- ZINC
- COBALT
- HYDROGEN
- POTASSIUM
- ANTIMONY
- FLUORINE
- MERCURY
- SILICON
- NICKEL

Spectrum Sudoku (p.144)

The spectrograms read 12456 (calcium), 12356 (magnesium), 12345 (caesium) and 12456 (calcium) – therefore calcium is the element that was detected twice.

7	9	5	6	3	1	2	8	4
6	4	8	5	9	2	1	3	7
2	1	3	8	7	4	9	6	5
1	2	4	3	5	9	8	7	6
3	6	7	2	1	8	4	5	9
5	8	9	4	6	7	3	2	1
4	5	6	9	8	3	7	1	2
9	3	1	7	2	5	6	4	8
8	7	2	1	4	6	5	9	3

Exoplanets (p.146)

The exoplanet names are:

- GALILEO: Named after Galileo Galilei, an Italian astronomer who was the first to use a telescope to see the skies
- BRAHE: Named after Tycho Brahe, a Danish astronomer famed for his accurate celestial observations

- ÆGIR: Named after the god of the sea in Old Norse mythology
- QUIJOTE: Named after the noble protagonist in a Spanish adventure novel, known in English as 'Don Quixote'
- POLTERGEIST: Named after a supernatural being whose name translates as 'noisy ghost'

Pulsar Puzzle (p.148)

The planets can be decoded as follows:

1. Venus
2. Jupiter
3. Mercury
4. Neptune
5. Mars
6. Saturn
7. Earth
8. Uranus

Cosmic Clouds (p.150)

The solution is as follows:

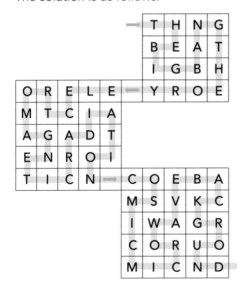

Therefore:

- The first grid reveals: THE BIG BANG THEORY
- The second grid reveals: ELECTROMAGNETIC RADIATION
- The third grid reveals: COSMIC MICROWAVE BACKGROUND

Black Hole Detector (p.152)

		2			2	2
3	4	3		2		
			1			
3	5		2			2
						3
3	5		4	4		
		2			3	2

Double Trouble (p.153)

	4		3		3	
		3		2		
2		1			8	
	2		3			4
1		2		7	7	4
	5		4			2
		4		4		2

Double Trouble Again (p.154)

The completed words, in the order they appear, are as follows:

- Hubble – American astronomer who gives their surname to an orbiting telescope
- apogee – point where the Moon's orbit is furthest from Earth
- ellipse – oval shape, like an orbit
- satellite – a planet-orbiting body
- gibbous – phase of the Moon

- vacuum – space without matter
- orrery – mechanical model of the Solar System
- annular – type of eclipse where the Sun remains visible as a ring around the Moon
- perigee – point where the Moon's orbit is closest to Earth
- full – phase of the Moon
- Halley – name of a comet, or its eponymous astronomer

The letters in each missing pair spell BELL BURNELL – the surname of Jocelyn Bell Burnell, the discoverer of pulsars.

Spaghettification (p.156)

In the order that the words appear in the puzzle, the solutions and their corresponding clues are as follows:
- SET = SATELLITE – an orbiting body
- GAS = GALAXIES – star systems
- SEA = SIDEREAL – relating to the stars
- PUB = PENUMBRA – area of partial shadow
- PAL = PARALLAX – apparent movement when a viewpoint changes
- LIE = ECLIPSED – obscured by a celestial body
- BAN = ABLATION – erosion caused by atmospheric friction

Galaxy Crossword (p.158)

The crossword can be filled in as follows:

The letters in the highlighted squares are M, R, A, D, N, O, A, D and E, which can be anagrammed to spell ANDROMEDA, the name of the closest galaxy to our own.

Spiral Galaxies (p.160)

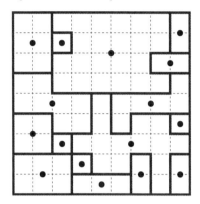

Star Search (p.162)

The names to find are as follows:
1. RING
2. ORION
3. HELIX
4. RED SQUARE
5. PELICAN
6. OWL
7. OYSTER
8. FOX FUR
9. ROSETTE
10. LAGOON
11. ANT
12. BOOMERANG

The nebulae can be found in the grid as follows:

Starry Start (p.164)

The names can be matched with the numbered images as follows:

1. Horsehead

2. Cat's Eye

3. Butterfly

4. Eagle

5. Tarantula

6. Crab

7. Stingray

CHAPTER 5:

NOW AND BEYOND

Space on Screen (p.170)

In the order given, the names of the films are:

- A Grand Day Out (1989)
- First Man (2018)
- 2001: A Space Odyssey (1968)
- Gravity (2013)
- Interstellar (2014)

- Hidden Figures (2016)
- Arrival (2016)
- The Martian (2015)
- Apollo 13 (1995)

Initially Heard (p.171)

In the order they appear, the titles are:

- *A Spaceman Came Travelling* – Chris de Burgh (song)
- *Rocket Man (I Think It's Going to Be a Long, Long Time)* – Elton John (song)
- *Starman* – David Bowie (song)
- *The Dark Side of the Moon* – Pink Floyd (album)
- *Jupiter* – Gustav Holst (orchestral movement)
- *Fly Me to the Moon* by Frank Sinatra (song)
- *Total Eclipse of the Heart* by Bonnie Tyler (song)
- *Here Comes the Sun* – The Beatles (song)

Jupiter, by Gustav Holst, is part of a set of seven orchestral pieces known as 'The Planets'. Each is themed after one of the planets in the Solar System, excluding Earth.

The answer to the musical bonus question is Brian May, guitarist of the band Queen. The band's lead singer was Freddie Mercury, who shared his surname with a planet.

Women and Space (p.172)

The full names of the women, in order, are:

- Valentina Tereshkova
- Svetlana Savitskaya
- Sally Ride
- Helen Sharman
- Caroline Herschel
- Katherine Johnson
- Maria Mitchell
- Jocelyn Bell Burnell

Their names can be traced in the grid as follows:

Doppler Effect (p.174)

In the order they appear in the puzzle, the astronomers are:
- (Albert) EINSTEIN
- (Stephen) HAWKING
- (Brian) COX
- (Neil deGrasse) TYSON
- (Edwin) HUBBLE
- (Carl) SAGAN
- (William) HARTMANN
- (Patrick) MOORE

High Flight (p.176)

The missing letters in the first verse are N, E, J, O, N, S, A and H:
Oh! I have slipped the surly bonds of Earth
And danced the skies on laughter-silvered wings;
Sunward I've climbed, and joined the tumbling mirth
of sun-split clouds, – and done a hundred things
You have not dreamed of – wheeled and soared and swung
High in the sunlit silence. Hov'ring there,
I've chased the shouting wind along, and flung
My eager craft through footless halls of air . . .

The missing letters in the second verse are L, E, K, R, P and E:
Up, up the long, delirious, burning blue
I've topped the wind-swept heights with easy grace.
Where never lark, or even eagle flew –
And, while with silent, lifting mind I've trod
The high untrespassed sanctity of space,
– Put out my hand, and touched the face of God.

The letters from the first verse can be rearranged to give
JOHANNES, and those from the second to give KEPLER,
revealing the German astronomer Johannes Kepler (1571–1630).

Space Agencies (p.178)

In the order they appear in the puzzle, the expanded abbreviations
and their respective countries or regions is as follows:
- CNSA – China National Space Administration – China
- ESA – European Space Agency – Europe
- ISRO – Indian Space Research Organisation – India
- JAXA – Japan Aerospace Exploration Agency – Japan
- NASA – National Aeronautics and Space Administration – USA

Two other agencies with full launch capacity are:
- RFSA: Roscosmos State Corporation for Space Activities –
  Russia's space agency, usually known just as Roscosmos
- CNES: French National Centre for Space Studies

One Small Step (p.179)

The hidden words are shown in **bold**:
1. IMP **AIR** FIELD = IMPAIR/AIRFIELD
2. FIRE **FLY** AWAY = FIREFLY/FLYAWAY
3. WONDER **LANDS** CAPES = WONDERLANDS/
   LANDSCAPES
4. FORT **NIGHT** FALL = FORTNIGHT/NIGHTFALL

254 ● ◐ ◯   THE ASTRONOMY PUZZLE BOOK

5. HONEY **MOON** BEAM = HONEYMOON/
MOONBEAM
LIME **LIGHT** WEIGHT = LIMELIGHT/LIGHTWEIGHT
The two words combine to create MOONLIGHT.
6. AIR **SPACE** CRAFT = AIRSPACE/SPACECRAFT
SLEEP **WALK** OUT = SLEEPWALK/WALKOUT
The two words combine to create SPACEWALK.
7. SUPER **STAR** BOARD = SUPERSTAR/STARBOARD
SAW **DUST** PAN = SAWDUST/DUSTPAN
The two words combine to create STARDUST.
8. AFTER **SUN** BATHE = AFTERSUN/SUNBATHE
HOT **SPOT** LESS = HOTSPOT/SPOTLESS
The two words combine to create SUNSPOT.

Scheduled Landing (p.180)

The rival's landing site is the one marked 'B' on the map. The
course of the flight path has been hidden using words that contain
'up', 'down', 'left' or 'right', which specify moves between adjacent
grid squares.

The hidden words are highlighted here:

Project E**up**horia is go. S**up**erb work so far and we have a b**right**
future ahead of us. This is going to be **downright** legendary. A
co**up**le of days and we'll be landing on the moon, barring any
hold-**up**s. The team has a month's worth of s**up**plies so there'll be
plenty **left** in case of emergency. Planning to land near the major
geological c**left**, but don't tell anyone else.

This gives a route of up, up, right, down, right, up, up, up, left, left,
which once converted to a path on the grid looks like this:

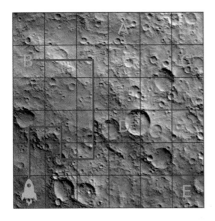

This reveals that the landing site is the one marked 'B'.

Space Food (p.182)

The food packages contain:

1. Beef and gravy
2. Strawberry cereal
3. Peaches
4. Beef sandwich

Spacecraft (p.184)

The combinations are as follows:
- CAB + OLUMI = COLUMBIA
- CAR + HLLENGE = CHALLENGER
- TUKTUK + SPNI I and SPNI II = SPUTNIK I and SPUTNIK II
- VAN + EDEOUR = ENDEAVOUR

The descriptions belong to the following spacecraft:

1. Sputnik I; Sputnik II
2. Endeavour
3. Columbia
4. Challenger

The Quadrangles of Mars (p.186)

The full names of the quadrangles of Mars whose names appear in the image are therefore:
- Amazonis
- Arabia
- Arcadia
- Elysium
- Eridania
- Hellas
- Memnonia
- Noachis
- Tharsis

Lost in Translation (p.188)

The symbols correspond with letters as follows:

A	C	E	G	I	J	N	O	OS	P	R	RI	S	T	U	V	Y
↪	⌐	↕	↨	⇉	‡	↳	↖	↕	↺	↻	↵	⌢	↗	↖	←	⚡

To double letter, add underline symbol

In the order they are given on the page, the coded names are therefore as follows:

CURIOSITY
CASSINI
ROSETTA
SPIRIT
JUNO
VOYAGER

For completeness, it should be noted that two Voyager probes were launched, *Voyager 1* and *Voyager 2*.

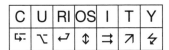

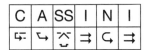

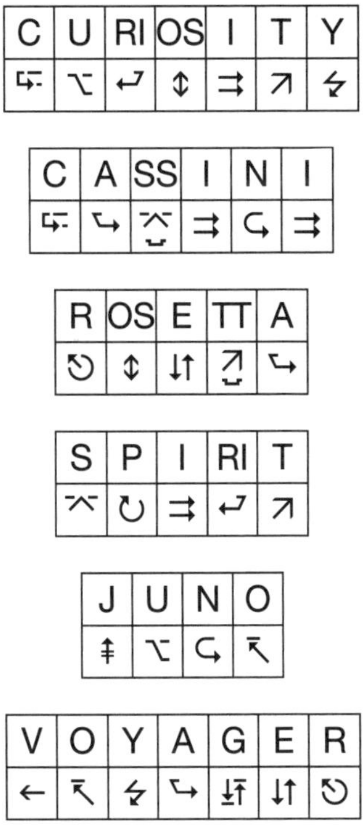

R	OS	E	TT	A
↻	↕	↓↑	↗	↳

S	P	I	RI	T
⌃	↻	⇉	↵	↗

J	U	N	O
�	⌐	↳	↰

V	O	Y	A	G	E	R
←	↰	↯	↳	↟	↓↑	↻

Time to Reflect (p.190)

The mirrors should be placed as shown below, creating the marked laser paths:

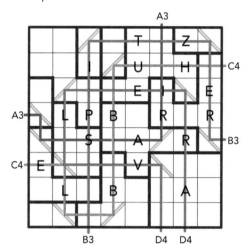

The names of the four telescopes highlighted by the beams are therefore:

- Hubble – internationally famous space telescope
- IRAS – the first space telescope to survey the night sky at infrared wavelengths
- Spitzer – followed on from IRAS and another telescope, ISO, as a dedicated infrared telescope
- Ariel V – former joint British and US telescope to observe X-rays

Fly-bys (p.192)

The planets shown are:
1. Jupiter
2. Mars
3. Saturn
4. Neptune

The image of Saturn shows its north pole, represented by the circular 'eye'.

Alien Announcement (p.194)

The 'X's are striking out letters from the previous transmission, while the 'O's are circling letters to keep. There is a direct letter-to-letter correspondence.

Combining 'Salutatory' with 'OXXXOOXXXO' therefore becomes 'S~alutator~y'– that is, 'Stay'.

Continuing similarly with each word in turn means that the whole message can then be decoded to the rather more menacing: 'Stay away from our planet'!

Martian Management (p.195)

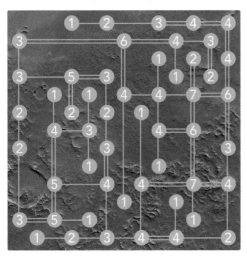

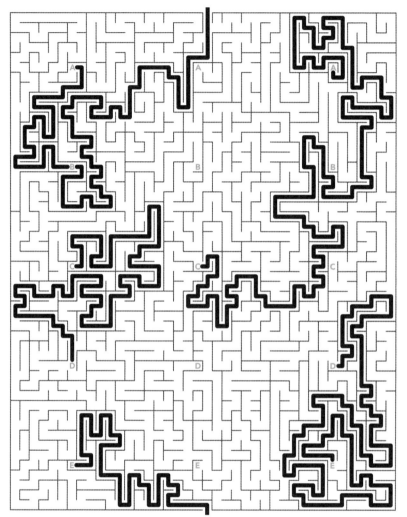

Wormhole Ways (p.196)

The Big Rip 1 (p.198)

```
|      | 2 |   5 |      |
|   6  |   8 |   9 |
| 8 3  |      |      |
|      | 4 |      | 5 |
|      |      |   3 |
|      | 3 | 3 |      |
|   8  |   4 | 8 |
|   6 | 8 |   4 |      |
|   3  |      |      |
```

The Big Rip 2 (p.199)

The restored names of the types of galaxies are as follows, in alphabetical order:

- DWARF
- ELLIPTICAL
- IRREGULAR
- LENTICULAR
- RING
- SHELL
- SPIRAL
- STARBURST

The galaxy named NGC 4151 is known as the 'Eye of Sauron' galaxy, after the antagonist in the fantasy book series.

Heat Death (p.200)

The restored words and their clues, in the order of the letter sets, are as follows:

- HOVERCRAFT – vehicle powered by downward jets of air
- SOUTH-EAST – location of Paris relative to Greenwich

- HELIOSTAT – astronomical equipment designed to reflect the sun's light in a given direction
- HUMECTANT – able to retain moisture
- HEAVIEST – uranium is the _____ naturally occurring chemical element
- THERMOSTAT – temperature-regulating device

The Big Crunch (p.202)

Each line contains two words for stretches of time, which have been 'crunched' together. In the order they appear in the puzzle, the two names in each line are:

- AGE and ERA
- YEAR and TERM
- AEON and SPAN
- DECADE and FORTNIGHT
- EPOCH and PERIOD
- CENTURY and MILLENNIUM
- CYCLE and PHASE

IMAGE PERMISSIONS

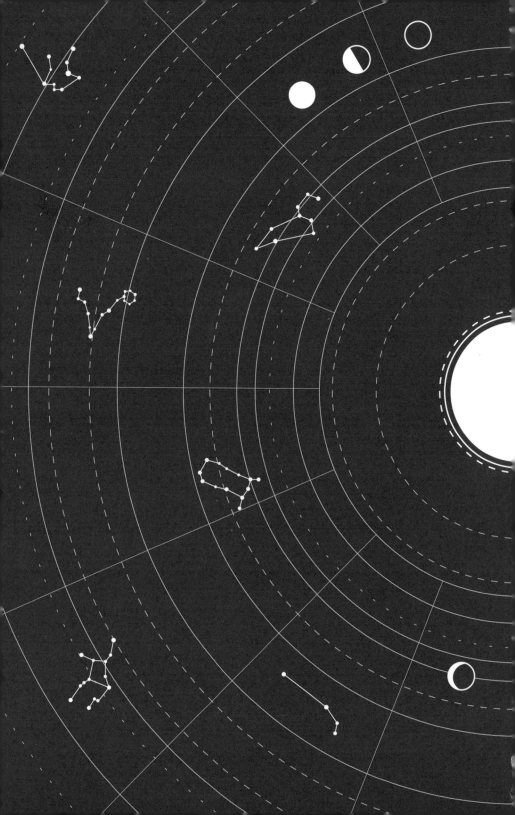